Sensing Emotions

Philips Research Book Series

VOLUME 12

Editor-in-Chief
Dr. Frank Toolenaar
Philips Research Laboratories, Eindhoven, The Netherlands

SCOPE TO THE '*PHILIPS RESEARCH BOOK SERIES*'

As one of the largest private sector research establishments in the world, Philips Research is shaping the future with innovative concepts that meet peoples' needs and desires. While the ultimate user benefits of these inventions end up on the high-street shelves, the often pioneering scientific and technological basis usually remains less visible.

This 'Philips Research Book Series' has been set up as a way for our researchers to contribute to the scientific community by publishing their comprehensive results and theories in a convenient book form.

Dr Peter Wierenga

Chief Executive Officer of Philips Research

For further volumes:
http://www.springer.com/series/6437

Sensing Emotions

The Impact of Context on Experience Measurements

Edited By

Joyce Westerink
Philips Research, Eindhoven, The Netherlands

Martijn Krans
Philips Research, Eindhoven, The Netherlands

and

Martin Ouwerkerk
Philips Research, Eindhoven, The Netherlands

Springer

Editors
Dr. Joyce Westerink
Philips Research
High Tech Campus 34
5656 AE Eindhoven
The Netherlands
joyce.westerink@philips.com

Dr. Martijn Krans
Philips Research
High Tech Campus 34
5656 AE Eindhoven
The Netherlands
martijn.krans@philips.com

Dr. Martin Ouwerkerk
Philips Research
High Tech Campus 34
5656 AE Eindhoven
The Netherlands
martin.ouwerkerk@philips.com

ISSN 1571-5671
ISBN 978-94-007-3712-9 ISBN 978-90-481-3258-4 (eBook)
DOI 10.1007/978-90-481-3258-4
Springer Dordrecht Heidelberg London New York

Softcover reprint of the hardcover 1st edition 2011

Printed on acid-free paper

Springer is part of Springer Science+Business Media (www.springer.com)

Introduction: Sensing Emotions in Context

We anticipate a future in which products and machines understand the mental state of people using them. A future in which products become more personalized by knowing how users feel and can adapt to the feelings they sense. Examples are music systems that effectively enhance our current mood with a personalized choice of music, computer dialogues that avoid upcoming frustration, and photo cameras that take pictures whenever we're excited. In all these situations, knowledge of the emotional state of the user is of prime importance.

A previous book published in the Philips Research Book Series, "Probing Experience", illustrated ways to evaluate the emotional state of the user through behavioral and physiological parameters. The majority of the authors were invited as speakers to the first "Probing Experience" symposium, held in Eindhoven, The Netherlands on June 7–8, 2006. As a sequel, on October 1, 2008, the international symposium "Probing Experience II" was held, also in Eindhoven, The Netherlands. The present book reflects the focus of this second "Probing Experience" symposium, highlighting the influence of context in these emotional state measurements. The authors of this book comprise world-leading researchers on this topic with a wide variety of backgrounds, from business and academia, and cover a broad range of context situations. Most of them contributed as speakers to "Probing Experience II".

The everyday-life contexts of future products and machines will always be specific, especially in comparison to the standard laboratory situation. Context can impact the experience measurements and influence the occurrence and characteristics of certain signals. On the other hand, independent knowledge of the context could be very valuable for the interpretation of experience measurements. Of course, the context situation is highly dependent on the application and user scenario of the product. In the various chapters, a broad range of user types is described ranging from athletes, people with obesity, problem sleepers, and pilots to post-traumatic stress disorder patients. The measurement techniques to determine the mental state of these users include interpretation of a variety of psychophysiological recordings (for instance heart rate, skin conductance, brain signals, muscle tension), as well as questionnaires, facial expression, and speech and text analysis.

The book opens with a chapter by Stephen Fairclough, in which he proposes a generic physiological computing paradigm, in which measurements of human physiology – since they reflect human experience – are fed into applications in real time.

Based on this input the applications can adapt their output so as to optimize this human experience. Though Fairclough's framework is confined to physiology, it can be extended to include other human-based inputs, like ratings, movement detection, and others.

Chapters 2 and 3 describe technology specifically developed to unobtrusively measure such parameters in real-life contexts. Martin Ouwerkerk presents the Emotion Measurement Platform, specifically its hardware, which measures skin conductance and heartbeats. Alberto Bonomi describes technology to derive information about the type and intensity of human activity from accelerometer measurements, and the requirements in different contexts.

Then three chapters follow that investigate examples in various specific contexts of how human experience is reflected in these human-based parameters. Wolfram Boucsein and co-authors show that several types of arousal are visible in the heartbeat and electrodermal parameters of a pilot during extreme flight maneuvers. In other contexts, however, this is not always the case, as Roos Rajae-Joordens describes in her chapter on the impact of colored room illumination. Emiel Krahmer and Marc Swerts illustrate how positive and negative experiences of speakers are reflected in their facial expressions as judged by others.

Having thus established that measurement technology is available for various contexts, and, what is more, is able to reflect human experience, we can start to think about application concepts dedicated to specific contexts. Four chapters illustrate the wide range of possibilities. Wayne Chase presents a number of communication enhancement tools that involve measurement of the emotional connotation of words. Joyce Westerink and co-authors evaluate the user experience with an application concept for runners that optimizes their workout based on their heartbeat. Henriette van Vugt tackles the domain of home sleep quality of healthy persons, and presents an overview of measurement and influencing techniques on which future concepts could be based. Egon van den Broek and co-authors target an application that improves the therapy of post-traumatic stress disorder patients, by indicating their most stressful moments as derived from several speech characteristics of their spoken words.

The book closes with a chapter of again a more generic nature, focusing on the process of developing such application concepts, which requires the cooperation of people with many backgrounds. Sharon Baurley describes her experience in the world of emotional fashion design, and proposes a design-based approach. Thus the Probing Experience II book follows the possibilities of applying human emotion sensing through a wide range of contexts – from a framework for application concepts, to a whole range of technologies, background research and application concepts, to a method of application concept generation. We sincerely hope you enjoy reading it.

Eindhoven, The Netherlands

Joyce Westerink
Martijn Krans
Martin Ouwerkerk

Contents

Contributors

Sharon Baurley School of Engineering and Design, Brunel University, Uxbridge, UK, sharon.baurley@brunel.ac.uk

Alberto G. Bonomi Medical Signal Processing Group, Philips Research, Eindhoven, The Netherlands, alberto.bonomi@philips.com

Wolfram Boucsein Physiological Psychology, University of Wuppertal, Wuppertal, Germany, boucsein@uni-wuppertal.de

Reinhard Braunstingl Department of Mechanics, Technical University Graz, Graz, Austria; Institute of Mechanics, Graz University of Technology, Vienna, Austria, r.braunstingl@TUGraz.at

Wayne O. Chase Connotative Intelligence Corporation, Vancouver, BC, Canada, wchase@connotative.com

Arjan Claassen Philips Innovation Services, Eindhoven, The Netherlands; Brain, Body and Behavior Group, Philips Research, Eindhoven, The Netherlands, Arjan.Claassen@philips.com

Yvonne de Kort Human Technology Interaction Group, Eindhoven University of Technology, Eindhoven, The Netherlands, Y.A.W.d.Kort@tue.nl

Ton Dijkstra Donders Institute for Brain, Cognition and Behavior, Radboud University Nijmegen, Nijmegen, The Netherlands, t.dijkstra@donders.ru.nl

Stephen H. Fairclough School of Natural Sciences and Psychology, Liverpool John Moores University, Liverpool, UK, s.fairclough@ljmu.ac.uk

Wijnand IJsselsteijn Human Technology Interaction Group, Eindhoven University of Technology, Eindhoven, The Netherlands, W.A.IJsselsteijn@tue.nl

K. Wolfgang Kallus Department of Work, Organizational and Environmental Psychology, Institute of Psychology, Karl-Franzens University of Graz, Graz, Austria, wolfgang.kallus@uni-graz.at

Ioana Koglbauer Brightline Avionics, Karlsdorf, Austria, i.koglbauer@gmail.com

Emiel Krahmer Faculty of Humanities, Tilburg centre for Cognition and Communication (TiCC), Tilburg University, Tilburg, The Netherlands, E.J.Krahmer@uvt.nl

Yuzhong Lin Architecture, Building and Planning, Eindhoven University of Technology, Eindhoven, The Netherlands; Brain, Body and Behavior Group, Philips Research, Eindhoven, The Netherlands, y.lin1@tue.nl

Martin Ouwerkerk Brain, Body and Behavior Group, Philips Research, Eindhoven, The Netherlands, martin.ouwerkerk@philips.com

Rosemarie J.E. Rajae-Joordens Philips Research, Eindhoven, The Netherlands, roos.rajae@philips.com

Tijn Schuurmans Soocial.com, Amsterdam, The Netherlands; Brain, Body and Behavior Group, Philips Research, Eindhoven, The Netherlands, tijn@tijnschuurmans.nl

Marc Swerts Faculty of Humanities, Tilburg centre for Cognition and Communication (TiCC), Tilburg University, Tilburg, The Netherlands, M.G.J.Swerts@uvt.nl

Guido van Helvoort Mindspark Interactive Network, White Plains, NY, USA; Brain, Body and Behavior Group, Philips Research, Eindhoven, The Netherlands, guido.vanhelvoort@mindspark.com

Henriette van Vugt Brain, Body and Behavior Group, Philips Research, Eindhoven, The Netherlands, henriette.van.vugt@philips.com

Egon L. van den Broek Human Media Interaction (HMI), University of Twente, Enschede, The Netherlands; Karakter University Center, Radboud University Medical Center Nijmegen, Nijmegen, The Netherlands, vandenbroek@acm.org

Frans van der Sluis Human Media Interaction (HMI), University of Twente, Enschede, The Netherlands, f.vandersluis@utwente.nl

Suzanne Vossen Advertisement, Uden, The Netherlands; Brain, Body and Behavior Group, Philips Research, Eindhoven, The Netherlands, suzanne@adversitement.nl

Joyce Westerink Brain, Body and Behavior Group, Philips Research, Eindhoven, The Netherlands, joyce.westerink@philips.com

Chapter 1
Physiological Computing: Interfacing with the Human Nervous System

Stephen H. Fairclough

Abstract This chapter describes the physiological computing paradigm where electrophysiological changes from the human nervous system are used to interface with a computer system in real time. Physiological computing systems are categorized into five categories: muscle interfaces, brain-computer interfaces, biofeedback, biocybernetic adaptation and ambulatory monitoring. The differences and similarities of each system are described. The chapter also discusses a number of fundamental issues for the design of physiological computing system, these include: the inference between physiology and behaviour, how the system represents behaviour, the concept of the biocybernetic control loop and ethical issues.

1.1 Introduction

Communication with computers is accomplished via a standard array of input devices requiring stereotypical actions such as key pressing, pointing and clicking. At the time of writing, the standard combination of keyboard/mouse is starting to yield to intuitive physical interfaces (Merrill and Maes, 2007), for instance, the Nintendo Wii and forthcoming "whole-body" interfaces such as Microsoft's Project Natal. Traditionally the physicality of human-computer interaction (HCI) has been subservient to the requirements of the input devices. This convention is currently in reversal as computers learn to understand the signs, symbols and gestures with which we physically express ourselves to other people. If users can communicate with technology using overt but natural hand gestures, the next step is for computers to recognise other forms of spontaneous human-human interaction, such as eye gaze (Wachsmuth et al., 2007), facial expressions (Bartlett et al., 2003) and postural changes (Ahn et al., 2007). These categories of expression involve subtle changes that are not always under conscious control. In one sense, these kinds of signals represent a more intuitive form of HCI compared to overt gesture because a person may communicate her needs to a device with very little intentionality. However, changes in facial expression or body posture remain overt and discernible by close visual

S.H. Fairclough (✉)
School of Natural Sciences and Psychology, Liverpool John Moores University, Liverpool, UK
e-mail: s.fairclough@ljmu.ac.uk

J. Westerink et al. (eds.), *Sensing Emotions*, Philips Research Book Series 12,
DOI 10.1007/978-90-481-3258-4_1, © Springer Science+Business Media B.V. 2011

observation. This progression of intuitive body interfaces reaches a natural conclusion when the user communicates with a computer system via physiological changes that occur under the skin. The body emits a wide array of bio-electrical signals, from increased muscle tension to changes in heart rate to tiny fluctuations in the electrical activity of the brain. These signals represent internal channels of communication between various components of human central nervous systems. These signals may also be used to infer behavioural states, such as exertion during exercise, but their real potential to innovate HCI lies in the ability of these measures to capture psychological processes and other dimensions that remain covert and imperceptible to the observer.

There is a long literature in the physiological computing tradition inspired by work on affective computing (Picard, 1997), specifically the use of psychophysiology to discern different emotional states and particularly those negative states such as frustration (Kapoor et al., 2007) that both designer and user wish to minimise or avoid. A parallel strand of human factors research (Pope et al., 1995; Prinzel et al., 2003) has focused on the detection of mental engagement using electroencephalographic (EEG) measures of brain activity. The context for this research is the development of safe and efficient cockpit automation; see Scerbo et al. (2003) for summary of automation work and Rani and Sarkar (2007) for similar approach to interaction with robots. The same approach was adopted to monitor the mental workload of an operator in order to avoid peaks (i.e. overload) that may jeopardise safe performance (Wilson and Russell, 2003, 2007). In these examples, psychophysiology is used to capture levels of cognitive processing rather than emotional states. Psychophysiology may also be used to quantify those motivational states underlying the experience of entertainment technology (Mandryk et al., 2006; Yannakakis et al., 2007). This application promotes the concept of adaptive computer games where software responds to the state of the player in order to challenge or help the individual as appropriate (Dekker and Champion, 2007; Fairclough, 2007; Gilleade and Dix, 2004). Specific changes in psychophysiology may also be used as an intentional input control to a computer system, Brain-Computer Interfaces (BCI) (Allison et al., 2007; Wolpaw et al., 2002) involve the production of volitional changes in EEG activity in order to direct a cursor and make selections in a manner similar to mouse movement or a key press.

Psychophysiology has the potential to quantify different psychological states (e.g. happiness vs. frustration), to index state changes along a psychological continuum (e.g. low vs. high frustration) and to function as a proxy for input control (e.g. a BCI). Psychophysiological data may also be used to identify stable personality traits, such as motivational tendencies (Coan and Allen, 2003) and predispositions related to health, such as stress reactivity (Cacioppo et al., 1998). The diversity and utility of psychophysiological monitoring provides ample opportunity to innovate HCI but what kinds of benefits will be delivered by a new generation of physiological computing systems? The first advantage is conceptual, contemporary human-computer communication has been described asymmetrical in the sense that the user can obtain a lot of information about the system (e.g. hard disk space, download speed, memory use) while the computer is essentially "blind" to the psychological status of the user (Hettinger et al., 2003). The physiological computing paradigm provides one route to a symmetrical HCI where both human and computer are capable of "reading" the status of the other without the requirement for the user to produce explicit cues; this symmetrical type of HCI can be described as a dialogue as opposed to the asymmetrical variety that corresponds to two monologues (Norman, 2007). One consequence of symmetrical HCI is that technology has the opportunity to demonstrate "intuition" or "intelligence" without any need to overtly consult the user. For example, a physiological computing system may offer help and advice based upon a psychophysiological diagnosis of frustration – or make a computer game more challenging if a state of boredom is detected. Given that the next generation of "smart" technology will be characterised by qualities such as increased autonomy and adaptive capability (Norman, 2007), future systems must be capable of responding proactively and implicitly to support human activity in the workplace and the home, e.g. ambient intelligence (Aarts, 2004). As technology develops in this direction, the interaction between users and machines will shift from a master-slave dyad towards the kind of collaborative, symbiotic relationship (Klein et al., 2004) that requires the computer to extend awareness of the user in real-time.

Each interaction between user and computer is unique at some level, the precise dynamic of the HCI is influenced by a wide range of variables originating from the individual user, the status of the system or the environment. The purpose of dialogue design is to create an optimal interface in order to maximise performance efficiency or safety, which represents a tacit attempt to "standardise" the dynamic of the HCI. Similarly, human factors and ergonomics research has focused on the optimisation of HCI for a generic "everyman" user. Physiological computing represents a challenge to the concepts of a standard interaction or a standard user. Interaction with a symmetrical physiological computing system incorporates a reflexive, improvisatory element as both user and system respond to feedback from the other in real-time. There may be benefits associated with this real-time, dynamic adaptation such as the process of individuation (Hancock et al., 2005) where the precise response of the system is tailored to the unique skills and preferences of each user, e.g. (Rashidi and Cook, 2009). As the individual develops an accurate model of system contingencies and competencies and vice versa, human-computer coordination should grow increasingly fluid and efficient. For example, certain parameters of the system (e.g. the interface) may change as the person develops from novice to experienced user, e.g. acting with greater autonomy, reducing the frequency of explicit feedback. This reciprocal human-machine coupling is characterised as a mutual process of co-evolution with similarities to the development of human-human relationships in teamwork (Klein et al., 2004). Central to this idealised interaction is the need to synchronise users' models of system functionality, performance characteristics etc. with the model of user generated by the computer system with respect to preferences, task context and task environment. In this way, physiological computing shifts the dynamic of the interaction from the generic to the specific attributes of the user. This shift is "directed to explore ways through which each and every individual can customize his or her tools to optimize the pleasure and efficiency of his or her personal interaction" (Hancock et al., 2005, p. 12).

Traditional input devices required a desktop space for keyboard or mouse that effectively tied HCI to a specific "office" environment. The advent of mobile communication devices and lightweight notebooks/laptops has freed the user from the desktop but not from the ubiquity of the keyboard or touchpad. The development of unintrusive, wearable sensors (Baber et al., 1999; Picard and Healey, 1997; Teller, 2004) offers an opportunity for users to communicate with ubiquitous technology without any overt input device. A psychophysiological representation of the user state could be collected unobtrusively and relayed to personal devices located on the person or elsewhere. Unobtrusive monitoring of physiology also provides a means for users to overtly communicate with computers whilst on the move or away from a desktop. The development of muscle-computer interfaces (Saponas et al., 2008) allows finger movements to be monitored and distinguished on potentially any surface in order to provide overt input to a device. Data collection from wearable sensors could be used to monitor health and develop telemedicine-related applications (Kosmack Vaara, Hook, and Tholander, 2009; Morris and Guilak, 2009) or to adapt technology in specific ways, e.g. if the user is asleep, switch all messages to voicemail. With respect to system adaptation, this "subconscious" HCI (i.e. when a device adapts to changes in user state without any awareness on the part of the user) could be very useful when the user is eyes- or hands-busy, such as driving a car or playing a computer game. This utilisation of the approach in this scenario allows physiological computing to extend the communication bandwidth of the user.

The potential benefits of physiological computing are counteracted by significant risks associated with the approach. The inference from physiological change to psychological state or behaviour or intention is not straightforward (Cacioppo et al., 2000a). Much of the work on the psycho-physiological inference (i.e. the way in which psychological significance is attached to patterns of physiological activity) has been conducted under controlled laboratory conditions and there is a question mark over the robustness of this inference in the field, i.e. psychophysiological changes may to be small and obscured by gross physical activity or environmental factors such as temperature. It is important that physiological computing applications are based upon a robust and reliable psychophysiological inference in order to work well. The physiological computing paradigm has the potential to greatly increase the complexity of the HCI which may be a risk in itself. If a physiological computing application adapts functionality or interface features in response to changes in the state of the user, this dynamic adaptation may be double-edged. It is hoped that this complexity may be harnessed to improve the quality of the HCI in terms of the degree of "intelligence" or "anticipation" exhibited by the system. However, the relationship between system complexity and compatibility with the user is often negative, i.e. the higher the complexity of the system, the lower the level of compatibility (Karwowski, 2000). Therefore, the complex interaction dynamic introduced by physiological computing devices has the potential to dramatically degrade system usability by increasing the degree of confusion or uncertainty on the part of the user. Finally, physiological computing approaches are designed to use physiology as a markers of what are often private, personal experiences. Physiological computing technologies cross the boundary between overt and covert expression, in some cases capturing subtle psychological changes of which the users may be unaware. This kind of technology represents a threat to privacy both in the sense of data security and in terms of feedback at the interface in a public space.

The aim of the current chapter is to describe different categories of physiological computing systems, to understand similarities and differences between each type of system, and to describe a series of fundamental issues that are relatively common to all varieties of physiological computing applications.

1.2 Categories of Physiological Computing

A physiological computing system is defined as a category of technology where electrophysiological data recorded directly from the central nervous system or muscle activity are used to interface with a computing device. This broad grouping covers a range of existing system concepts, such as Brain-Computer Interfaces (Allison et al., 2007), affective computing (Picard, 1997) and ambulatory monitoring (Ebner-Priemer and Trill, 2009). This definition excludes systems that classify behavioural change based on automated analysis of gestures, posture, facial expression or vocal characteristics. In some cases, this distinction merely refers to the method of measurement rather than the data points themselves; for example, vertical and horizontal eye movement may be measured directly from the musculature of the eye via the electrooculogram (EOG) or detected remotely via eye monitoring technology where x and y coordinates of gaze position are inferred from tracking the movement of pupil.

Figure 1.1 (below) describes a range of physiological computing systems that are compared and contrasted with overt input control derived from conventional keyboard/mouse or gesture-based control [1]. The second category of technology describes those physiological computing concepts where input control is based upon

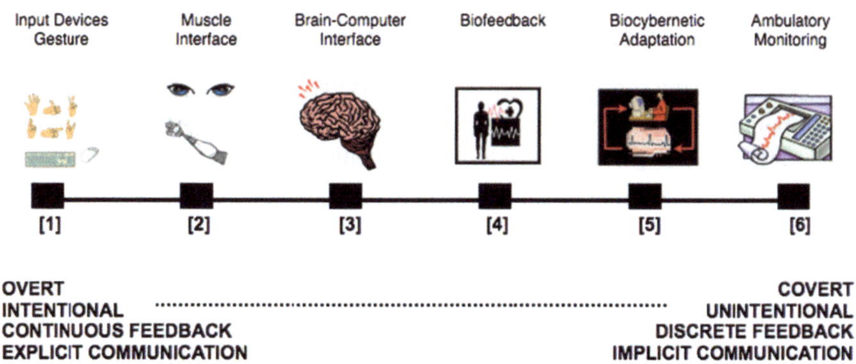

Fig. 1.1 Five categories of physiological computing systems

muscular activity [2]. These systems include cursor control using eye movements (Tecce et al., 1998) or gaze monitoring (Chin et al., 2008) or eye blink activity (Grauman et al., 2001). Muscle interfaces have traditionally been explored to offer alternative means of input control for the people with disabilities and the elderly (Murata, 2006). The same "muscle-interface" approach using electromyographic (EMG) activity has been used to capture different hand gestures by monitoring the muscles of the forearm (Saponas et al., 2008), facial expressions (Huang et al., 2006) and subvocal speech (Naik et al., 2008). Brain-Computer Interfaces (BCI) [3] are perhaps the best known variety of physiological computing system. These systems were originally developed for users with profound disabilities (Allison et al., 2007; Wolpaw et al., 2002) and indexed significant changes in the electrical activity of the cortex via the electroencephalogram (EEG), e.g. evoked-potentials (ERPs), steady state visual evoked potentials (SSVEPs). Several arguments have been forwarded to promote the use of BCI by healthy users (Allison et al., 2007), such as novelty or to offer an alternative mode of input for the "hands-busy" operator. Zander and Jatzev (2009) distinguished between active BCI systems that rely on direct EEG correlates of intended action (e.g. changes in the somatosensory cortex in response to motor imagery) and reactive BCI where EEG activity is not directly associated with output control (e.g. use of P300 ERP amplitude to a flashing array of letters to enable alphanumeric input). Biofeedback systems [4] represent the oldest form of physiological computing. The purpose of this technology is to represent the physiological activity of the body in order to promote improved self-regulation (Schwartz and Andrasik, 2003). This approach has been applied to a range of conditions, such as asthma, migraines, attentional deficit disorder and as relaxation therapy to treat anxiety-related disorders and hypertension. Biofeedback therapies are based on monitoring the cardiovascular system (e.g. heart rate, blood pressure), respiratory variables (e.g. breathing rate, depth of respiration), EMG activity, and EEG (i.e. neurofeedback) and training users to develop a degree of volitional control over displayed physiological activity. The concept of biocybernetic adaptation [5] was developed by Pope et al. (1995) to describe a adaptive computer system that responded to changes in EEG activity by controlling provision of system automation (Freeman et al., 2004; Prinzel et al., 2003). This types of system monitor naturalistic changes in the psychological state of the person, which may be related to variations in cognitive workload (Wilson and Russell, 2003) or motivation and emotion (Mandryk and Atkins, 2007; Picard et al., 2001). This approach has been termed "wiretapping" (Wolpaw et al., 2000) or passive BCI (Zander and Jatzev, 2009). In essence, the psychological status of the user is monitored in order to trigger software adaptation that is both timely and intuitive (Fairclough, 2009). The final category of technology concerns the use of unobtrusive wearable sensors that monitor physiological activity over a sustained period of days or months. These ambulatory systems [6] may be used to monitor emotional changes (Picard and Healey, 1997; Teller, 2004) or health-related variables (McFetridge-Durdle et al., 2008; Milenkovic et al., 2006). These systems may trigger feedback to the individual from a mobile device when "unhealthy" changes are detected (Morris, 2007) or the person may review personal data on a retrospective basis (Kosmack Vaara et al., 2009).

The biocybernetic loop is a core concept for all physiological computing systems (Fairclough and Venables, 2004; Pope et al., 1995; Prinzel et al., 2000) with the exception of some forms of ambulatory monitoring [6]. This loop corresponds to a basic translational module that transforms physiological data into a form of computer control input in real-time. The loop has at least three distinct stages: (1) signal acquisition, filtering and digitization, (2) artifact correction and the extraction of relevant features and (3) the translation of an attenuated signal into output for computer control. The precise form of the mapping between physiological change and control output will differ from system to system; in some cases, it is relatively literal and representative, e.g. the relationship between eye movements and x,y coordinates in space. Other systems involve a symbolic mapping where physiological activity is converted into a categorization scheme that has psychological meaning. For example, the relationship between autonomic activity and emotional states falls into this category (Mandryk and Atkins, 2007), similarly the mapping between EEG activity and mental workload (Gevins et al., 1998; Grimes et al., 2008) or the way in which respiratory data may be represented as sound or visual animation via a biofeedback interface. These mappings have been developed primarily to produce one-dimensional output, although there are two-dimensional examples of both BCI (Wolpaw and McFarland, 2004) and biocybernetic adaptation (Rani et al., 2002). Sensitivity gradation is a common issue for many biocybernetic loops. Some forms of BCI and all forms of biocybernetic adaptation rely on an attenuated signal for output, for example, a steady and gradual increase over a specified time window. In the case of ambulatory monitoring, some systems alert the user to "unhealthy" physiological activity use the same kind of sensitivity gradation to trigger an alert or diagnosis. Those ambulatory systems that do not incorporate a biocybernetic loop are those that rely exclusively on retrospective data, such as the affective diary concept (Kosmack Vaara et al., 2009); in this case, real time data is simply acquired, digitised, analysed and conveyed to the user in various formats without any translation into computer control.

The five categories of physiological computing system illustrated in Fig. 1.1 have been arranged to emphasise important differences and similarities. Like conventional input via keyboard and mouse, it is argued that muscle interfaces involving gestures, facial expressions or eye movements are relatively overt and visible to an observer. The remaining systems to the right of the diagram communicate with computer technology via covert changes in physiological activity. When a user communicates with a computer via keyboard/mouse [1], muscle interface [2] or BCI [3], we assume these inputs are intentional in the sense that the user wishes to achieve a specific action. The use of a Biofeedback system [4] is also volitional in the sense that the person uses the interface in order to manipulate or self-regulate a physiological response. By contrast, Biocybernetic Adaptation [5] involves monitoring spontaneous physiological activity in order to represent the state of the user with respect to a specific psychological dimension, such as emotion or cognitive workload. This is an unintentional process during which the user essentially remains passive (Fairclough, 2007, 2008). The same is true of ambulatory monitoring systems [6] that conform to the same dynamic of user passivity. Muscle Interfaces [2],

BCIs [3] and biofeedback [4] all operate with continuous feedback. Both Muscle Interfaces and BCIs are analogous to command inputs such as keystrokes, discrete gestures or mouse movements; these devices require continuous feedback in order to function. Biofeedback systems also rely on continuous feedback to provide users with the high-fidelity of information necessary to manipulate the activity of the central nervous system. In this case, the computer interface is simply a conduit that displays physiological activity in an accessible form for the user. Those physiological computing systems described as Biocybernetic Adaptation [5] rely on a different dynamic where feedback may be presented in a discrete form. For example, adaptive automation systems may signal a consistent trend, such as increased task engagement over a period of seconds or minutes, by activating an auto-pilot facility (Prinzel et al., 2002); similarly, a computer learning environment could signal the detection of frustration by offering help or assistance to the user (Burleson and Picard, 2004; Gilleade et al., 2005). The contingencies underlying this discrete feedback may not always be transparent to the user; in addition, discrete feedback may be delayed in the sense that it represents a retrospective trend. Ambulatory Monitoring systems [6] are capable of delivering relatively instant feedback or reflecting a data log of hours or days. In the case of ambulatory systems, much depends on why these data are recorded. Ambulatory recording for personal use tends to fall into two categories: (1) quantifying physiological activity during specific activities such as jogging and (2) capturing physiological activity for diary or journal purposes. In the case of the former, feedback is delivered in high fidelity (e.g. one reading every 15 or 30 s), whereas journal monitoring may aggregate data over longer time windows (e.g. one reading per hour).

The biocybernetic control loop serves a distinct purpose when physiology is used as an explicit channel for communication with a computing device, e.g. muscle interface [2], BCI [3]. In these cases, physiological activity is translated into analogues of distinct actions, to activate a function or identify a letter or move a cursor through two-dimensional space. Biocybernetic Adaptation [5] is designed to mediate an implicit interaction between the status of the user and the meta-goals of the HCI (Fairclough, 2008). The latter refers to the design goals of the technological device; in the case of an adaptive automation system, the meta-goals are to promote safe and efficient performance; for a computer game, the goal would be to entertain and engage. Biocybernetic Adaptation [5] provides the opportunity for real-time adjustment during each interaction in order to reinforce the design goals of the technology. Finally, there may be a requirement for training when physiology is used as a means of explicitly computer control. Muscle-based interaction [2] may require some familiarisation as user adjust to the sensitivity of system response. BCI devices [3] are often associated with a training regime, although there is evidence that their training requirements may not be particularly onerous (Guger et al., 2003). Biofeedback systems [4] are designed as a training tool for self-regulation. However, physiological computing systems that rely on implicit communication such as Biocybernetic Adaptation [5] and Ambulatory Monitoring [6] have no training requirement from the perspective of the user.

The continuum of physiological computing systems illustrated in Fig. 1.1 obscures the huge overlap between different categories. Ambulatory monitoring [6] represents a common denominator for all other physiological computing systems, i.e. if a system records electrophysiological activity from the user, these data can also be used for affective diaries or health monitoring. In addition, it is anticipated that wearable sensors currently associated with ambulatory monitoring will become the norm for all physiological computing systems in the future. A biofeedback component [4] is also ubiquitous across all systems. Users of Muscle Interfaces [2] and BCIs [3] rely on feedback at the interface in order to train themselves to produce reliable gestures or consistent changes in EEG activity. In these cases, the success or failure of a desired input control represents a mode of biofeedback. Biocybernetic Adaptation [5] may also include an element of biofeedback; these systems monitor implicit changes in psychophysiology in order to adapt the interface, but if these adaptations are explicit and consistently associated with distinct physiological changes, then changes at the interface will function as a form of biofeedback. Furthermore, if the user of a Biocybernetic Adaptation system [5] learns how to self-regulate physiology via biofeedback [4], this opens up the possibility of volitional control (over physiology) to directly and intentionally control system adaptation; in this case, the Biocybernetic Adaptation system [5] may be operated in the overt, intentional mode normally used to characterise Muscle Interfaces [2] and BCI [3]. There are a number of system concepts already available that combine Ambulatory Monitoring [6] with Biofeedback [4]; for instance, the Home Heart system (Morris, 2007) that monitors stress-related cardiovascular changes and triggers a biofeedback exercise as a stress countermeasure.

By breaking down the distinction between different types of physiological computing system in Fig. 1.1, we may also consider hybrid systems that blend different modes of input control and system adaptation. For example, it is difficult to imagine BCI technology being attractive to healthy users because of its limited bandwidth, e.g. two degree of spatial freedom, or two-choice direct selection. A hybrid system where BCI is used alongside a keyboard, mouse or console appears a more likely option, but the design of such a system faces two primary obstacles (Allison et al., 2007): (1) assigning functionality to the BCI that is intuitive, complimentary and compatible with other input devices, and (2) limitations on human information processing in a multi-tasking framework. The multiple-resource model (Wickens, 2002) predicts that control via BCI may distract attention from other input activities via two routes: sharing the same processing code (spatial vs. verbal) or by demanding attention at an executive or central level of processing. However, there is evidence that these types of time-sharing deficits may be overcome by training (Allison et al., 2007). The combination of Muscle Interfaces and BCI may work well for hands-free locate-and-select activities such as choosing from an array of images; eye movement may be used to locate the desired location in space and a discrete BCI trigger from the EEG used to make a selection. Biocybernetic Adaptation may be combined with either Muscle Interfaces or BCI because the former operate at a different level of the HCI (Fairclough, 2008). A system that trained users how to operate a Muscle Interface or a BCI could incorporate a biocybernetic adaptive element whereby the system offered help or advice based on the level of stress or workload associated with the training programme. Similarly, Biocybernetic Adaptation may be combined with conventional controls or gesture input to operate as an additional channel of communication between user and system. Those physiological computing systems such as Biocybernetic Adaptation or Ambulatory Monitoring that emphasise monitoring of behavioural states could also be combined with sensors that detect overt changes in facial expression, posture or vocal characteristics to create a multi-modal representation of the user, e.g. Kapoor et al. (2007).

Physiological computing systems may be described along a continuum from overt and intentional input control with continuous feedback to covert and passive monitoring systems that provide feedback on a discrete basis. There is a large overlap between distinct categories of physiological computing systems and enormous potential to use combinations or hybrid versions.

1.3 Fundamental Issues

The development of physiological computing remains at an early stage and research efforts converge on several fundamental issues. The purpose of this section is to articulate issues that have a critical bearing on the development and evaluation of physiological computing systems.

1.3.1 The Psychophysiological Inference

The complexity of the psychophysiological inference (Cacioppo and Tassinary, 1990; Cacioppo et al., 2000b) represents a significant obstacle for the design of physiological computing systems. The rationale of the biocybernetic control loop is based on the assumption that the psychophysiological measure (or array of measures) is an accurate representation of a relevant psychological element or dimension, e.g. hand movement, frustration, task engagement. This assumption is often problematic because the relationship between physiology and psychology is inherently complex. Cacioppo and colleagues (1990; 2000) described four possible categories of relationship between physiological measures and psychological elements:

- One-to-one (i.e. a physiological variable has a unique isomorphic relationship with a psychological or behavioural element)
- Many-to-one (i.e. two or more physiological variables are associated with the relevant psychological or behavioural element)
- One-to-many (i.e. a physiological variable is sensitive to one or more psychological or behavioural elements)
- Many-to-many (i.e. several physiological variables is associated with several psychological or behavioural elements)

The implications of this analysis for the design of physiological computing systems should be clear. The one-to-many or many-to-many categories that dominate the research literature represent psycho-physiological links that are neither exclusive nor uncontaminated. This quality is captured by the diagnosticity of the psychophysiological measure, i.e. the ability of the measure to target a specific psychological concept or behaviour and remain unaffected by related influences (O'Donnell and Eggemeier, 1986). In the case of Muscle Interfaces, it is assumed that one-to-one mapping between physiology and desired output may be relatively easy to obtain, e.g. move eyes upwards to move cursor in desired direction. For other systems such as BCI and particularly biocybernetic adaptation, finding a psychophysiological inference that is sufficiently diagnostic may be more problematic. Whilst it is important to maximise the diagnosticity of those measures underlying a physiological computing system, it is difficult to translate this general requirement into a specific guideline. Levels of diagnostic fidelity will vary for different systems. The system designer must establish the acceptable level of diagnosticity within the specific context of the task and the system.

1.3.2 The Representation of Behaviour

Once psychophysiological inference has been established, the designer may consider how specific forms of reactivity (e.g. muscle tension, ERPs) and changes in the psychological state of the user should be operationalised by the system. This is an important aspect of system design that determines:

- the transfer dynamic of how changes in muscle tension translate into movement of a cursor for a muscle interface
- the relationship between activity in the sensorimotor cortex and output to wheelchair control for a BCI
- the relationship between changes in EEG and autonomic activity and the triggering of adaptive strategies during biocybernetic adaptation

The biocybernetic loop encompasses the decision-making process underlying software adaptation. In its simplest form, these decision-making rules may be expressed as simple Boolean statements; for example, IF frustration is detected THEN offer help. The loop incorporates not only the decision-making rules, but in the case of Biocybernetic Adaptation, the psychophysiological inference implicit in the quantification of those trigger points used to activate the rules. In our study (Fairclough et al., 2006) for example, this information took the form of a linear equation to represent the state of the user, e.g. subjective mental effort = x_1 * respiration rate – x_2 * eye blink frequency + intercept, as well as the quantification of trigger points, e.g. IF subjective effort > y THEN adapt system. Other studies have also used linear modelling techniques and more sophisticated machine learning approaches systems to characterise user state in terms of the psychophysiological response, e.g. (Liu et al., 2005; Mandryk and Atkins, 2007; Rani et al., 2002; Wilson and Russell, 2003).

The psychological state of the user has been represented as a one-dimensional continuum, e.g. frustration (Gilleade and Dix, 2004; Kapoor et al., 2007; Scheirer et al., 2002), anxiety (Rani et al., 2005), task engagement (Prinzel et al., 2000), mental workload (Wilson and Russell, 2007). Other research has elected to represent user state in terms of: distinct categories of emotion (Healey and Picard, 1997; Lisetti and Nasoz, 2004; Lisetti et al., 2003), two-dimensional space of activation and valence (Kulic and Croft, 2005, 2006) and distinct emotional categories based upon a two-dimensional analysis of activation and valence (Mandryk and Atkins, 2007) As stated earlier, reliance on a one-dimensional representation of the user may restrict the range of adaptive options available to the system. This may not be a problem for some systems, but complex adaptation requires a more elaborated representation of the user in order to extend the repertoire of adaptive responses.

Early examples of physiological computer systems will rely on one-dimensional representations of the user, capable of relatively simple adaptive responses. The full potential of the technology may only be realized when systems are capable of drawing from an extended repertoire of precise adaptations, which will require complex representations of user behaviour or state in order to function.

1.3.3 The Biocybernetic Control Loop

The design of a physiological computing system is based upon the biocybernetic control loop (Fairclough and Venables, 2004; Pope et al., 1995; Prinzel et al., 2000). The biocybernetic loop defines the modus operandi of the system and is represented as a series of contingencies between psychophysiological reactivity and system responses or adaptation. These rules are formulated to serve a meta-goal or series of meta-goals to provide the system with a tangible and objective rationale. The meta-goals of the biocybernetic loop must be carefully defined and operationalised to embody generalised human values that protect and enfranchise the user (Hancock, 1996). For example, the physiological computing system may serve a preventative meta-goal, i.e. to minimise any risks to the health or safety of the operator and other persons. Alternatively, meta-goals may be defined in a positive way that promotes pleasurable HCI (Hancock et al., 2005; Helander and Tham, 2003) or states of active engagement assumed to be beneficial for both performance and personal well-being.

The biocybernetic loop is equipped with a repertoire of behavioural responses or adaptive interventions to promote the meta-goals of the system, e.g. to provide help, to give emotional support, to manipulate task difficulty (Gilleade et al., 2005). The implementation of these interventions is controlled by the loop in order to "manage" the psychological state of the user. Correspondingly, the way in which person responds to each adaptation is how the user "manages" the biocybernetic loop. This is the improvisatory crux that achieves human-computer collaboration by having person and machine respond dynamically and reactively to responses from each other. It may be useful for the loop to monitor how users respond to each intervention in order to individualise (Hancock et al., 2005) and refine this dialogue. This generative and recursive model of HCI emphasises the importance of: (a) accurately monitoring the psychological state of the user (as discussed in the previous sections), and (b) equipping software with a repertoire of adaptive responses that covers the full range of possible outcomes within the human-computer dialogue over a period of sustained use. The latter point is particularly important for "future-proofing" the physiological computing system as user and machine are locked into a co-evolutionary spiral of mutual adaptation (Fairclough, 2007).

Research into motivation for players of computer games has emphasised the importance of autonomy and competence (Ryan et al., 2006), i.e. choice of action, challenge and the opportunity to acquire new skills. This kind of finding begs the question of whether the introduction of a biocybernetic loop, which "manages" the HCI according to preconceived meta-goals, represents a threat to the autonomy and competence of the user? Software designed to automatically help or manipulate task demand runs the risk of disempowerment by preventing excessive exposure to either success or failure. This problem was articulated by Picard and Klein (2002) who used the phrase "computational soma" to describe affective computing software that effectively diffused and neutralised negative emotions. Feelings of frustration or anger serve as potent motivators within the context of a learning process; similarly, anxiety or fatigue are valuable psychological cues for the operator of a safety-critical system. It is important that the sensitivity of the biocybernetic loop is engineered to prevent over-corrective activation and interventions are made according to a conservative regime. In other words, the user should be allowed to experience a negative emotional state before the system responds. This is necessary for the system to demonstrate face validity, but not to constrain users' self-regulation of behaviour and mood to an excessive degree.

The biocybernetic loop encapsulates the values of the system and embodies a dynamic that promotes stable or unstable task performance. The dynamics of the control loop may be alternated for certain application to avoid the placement of excessive constraints on user behaviour.

1.4 Ethics and Privacy

A number of ethical issues are associated with the design and use of physiological computing systems. This technology is designed to tap private psychophysiological events and use these data as the operational fulcrum for a dynamic HCI. The ethical intention and values of the system designer are expressed by the meta-goals that control the biocybernetic loop (see previous section), but regardless of designers' good intentions, the design of any technology may be subverted to undesirable ends and physiological computing systems offer a number of possibilities for abuse (Reynolds and Picard, 2005b).

Invasion of privacy is one area of crucial concern for users of physiological computing systems. Ironically, a technology designed to promote symmetrical communication between user and system creates significant potential for asymmetry with respect to data protection, i.e. the system may not tell the user where his or her data are stored and who has access to these data. If data protection rights are honored by the physiological computing system, it follows that ownership of psychophysiological data should be retained formally and legally by the individual (Hancock and Szalma, 2003). One's own psychophysiological data are potentially very sensitive and access to other parties and outside agencies should be subject to formal consent from the user; certain categories of psychophysiological data may be used to detect medical conditions (e.g. cardiac arrhythmias, hypertension, epilepsy) of which the individual may not even be aware. The introduction of physiological computing should not provide a covert means of monitoring individuals for routine health problems without consent. In a similar vein, Picard and Klein (2002) argued that control of the monitoring function used by an affective computing system should always lie with the user. This is laudable but impractical for the user who wishes to benefit from physiological computing technology whilst enjoying private data collection. However, granting the user full control over the mechanics of the data collection process is an important means of reinforcing trust in the system.

Kelly (2006) proposed four criteria for information exchange between surveillance systems and users that are relevant here:

1. The user knows exactly what information is being collected, why it is being collected, where these data are stored and who has access to these data.
2. The user has provided explicit or implicit consent for data collection and can demonstrate full knowledge of data collection.
3. The user has access to these data, the user may edit these data or use these data himself or herself
4. Users receive some benefit for allowing the system to collect these data (e.g. recommendations, filtering).

This "open source" relationship between user and technology is called reciprocal accountability (Brin, 1999). This relationship may be acceptable for users of physiological computing systems provided the apparent transparency of the process does not mask crucial inequalities, i.e. vague formulations of data rights by private companies or governments. The provision of written consent to specify this relationship should allay users' concerns and there is evidence (Reynolds and Picard, 2005a) to support this position.

A second threat to privacy concerns how psychophysiological data recorded in real-time may be expressed at the interface, i.e. feedback at the interface on user state may be perceived by colleagues or other persons when the computer is situated in a public space. The provision of explicit verbal messages or discrete text/symbolic messages in response to the detection of frustration or boredom are potentially embarrassing for the user in the presence of others. The fact that computer systems are used in public spaces constitutes a call for discretion on the part of the interface design, particularly with respect to the use of auditory feedback. It would also be essential to include a facility that enables users to disable those messages or modes of feedback that leave them susceptible to "eavesdropping" by others.

Physiological computing systems are designed to "manipulate" the state of the user in a benign direction via the positive meta-goals of the biocybernetic loop. But how do users feel about being manipulated by autonomous technology (Picard and Klein, 2002; Reynolds and Picard, 2005a)? The verb "manipulate" is a loaded term in this context as people manipulate their psychological state routinely via psychoactive agents (e.g. caffeine, nicotine, alcohol), leisure activities (e.g. exercise, playing computer games) and aesthetic pastimes (e.g. listening to music, watching a TV show or movie) (Picard and Klein, 2002). The issue here is not the manipulation of psychological state per se but rather who retains control over the process of manipulation. When a person exercises or listens to music, they have full control over the duration or intensity of the experience, and may balk at the prospect of ceding any degree of control to autonomous technology. These concerns reinforce arguments that reciprocal accountability and granting the individual full control over the system are essential strategies to both reassure and protect the user. In addition, users need to understand how the system works so they are able understand the range of manipulations they may be subjected to, i.e. an analytic method for tuning trust in an automated system (Miller, 2005).

Physiological computing systems have the potential to be subverted to achieve undesirable outcomes such as invasion of privacy and tacit manipulation of the user. It is impossible to safeguard any new technology in this respect but provision of full transparency and reciprocal accountability drastically reduces the potential for abuse. It is important that the user of a physiological computing system remains in full control of the process of data collection (Picard and Klein, 2002) as this category of autonomous technology must be designed to empower the user at every opportunity (Hancock, 1996; Norman, 2007).

1.5 Summary

The concept of physiological computing allows computer technology to interface directly with the human nervous system. This innovation will allow users to provide direct input control to technology via specific changes in muscle tension and brain activity that are intentional. Data provided by wearable sensors can be used to drive biocybernetic adaptation and for ambulatory monitoring of physiological activity. In these cases, physiological changes are passively monitored and used as drivers of real-time system adaptation (biocybernetic adaptation) or to mark specific patterns that have consequences for health (ambulatory monitoring). The concept of biofeedback is fundamental to all categories of physiological computing as users may use these systems to promote increased self-regulation with respect to novel input devices (muscle interfaces or BCI), emotional control and stress management. Five different categories of physiological computing systems have been described (Muscle Interface, BCI, Biofeedback, Biocybernetic Adaptation, Ambulatory Monitoring) and there is significant overlap between each category. In addition, these physiological computing systems may be used to augment conventional input control in order to extend the communication bandwidth of the HCI.

The benefits of the physiological computing paradigm are counteracted by a number of potential risks, including systems that provide a mismatch with the behavioural state of the user or diminish user autonomy or represent a considerable threat to personal privacy. It is argued that the sensitivity of physiological computing system is determined by the diagnosticity of the psycho-physiological inference, i.e. the ability of the physiological data to consistently index target behaviour regardless of environmental factors or individual differences. It was also proposed that the biocybernetic control loop (the process by which physiological changes are translated into computer control) be carefully designed in order to promote design goals (e.g. safety and efficiency) without jeopardising the primacy of user control. The privacy of the individual is of paramount importance if physiological computing systems are to be acceptable to the public at large. A number of security issues were discussed with reference to controlling access to personal data and empowering the data protection rights of the individual.

References

Aarts, E.: Ambient intelligence: a multimedia perspective. IEEE Multimed. **11**(1), 12–19 (2004)

Ahn, H., Teeters, A., Wang, A, Breazeal, C., Picard, R.W.: Stoop to conquer: posture and affect interact to influence computer users' persistence. Paper presented at the second international conference on affective computing and intelligent interaction, Lisbon, Portugal (2007)

Allison, B., Graimann, B., Graser, A.: Why use a BCI if you are healthy? Paper presented at the ACE workshop – Brainplay'07: brain-computer interfaces and games, Salzburg (2007)

Allison, B.Z., Wolpaw, E.W., Wolpaw, J.R.: Brain-computer interface systems: progress and prospects. Expert Rev. Med. Devices **4**(4), 463–474 (2007)

Baber, C., Haniff, D.J., Woolley, S.I.: Contrasting paradigms for the development of wearable computers. IBM Syst. J. **38**(4), 551–565 (1999)

Bartlett, M., Littlewort, G., Fasel, I., Morvellan, J.R.: Real time face detection and facial expression recognition: development and application to human-computer interaction. Paper presented at the computer vision and pattern recognition for human-computer interaction, Vancouver (2003)

Brin, D.: The Transparent Society: Will Technology Force Us to Choose between Privacy and Freedom. Perseus Books, New York, NY (1999)

Burleson, W., Picard, R.W.: Affective agents: sustaining motivation to learn through failure and a state of "stuck". Paper presented at the workshop on social and emotional intelligence in learning environments, Maceio, Alagoas, Brazil (2004)

Cacioppo, J.T., Berntson, G.G., Malarkey, W., Kiecolt-Glaser, J.K., Sheridan, J.F., Poehlmann, K. et al.: Autonomic, neuroendocrine and immune responses to psychological stress: the reactivity hypothesis. Ann. N Y Acad. Sci. **840**, 664–673 (1998)

Cacioppo, J.T., Tassinary, L.G.: Inferring psychological significance from physiological signals. Am. Psychol. **45**(1), 16–28 (1990)

Cacioppo, J.T., Tassinary, L.G., Berntson, G.G.: Handbook of Psychophysiology. Cambridge University Press, New York, NY (2000a)

Cacioppo, J.T., Tassinary, L.G., Berntson, G.G.: Psychophysiological science. In: Cacioppo, J.T., Tassinary, L.G., Berntson, G.G. (eds.) Handbook of Psychophysiology, pp. 3–26. Cambridge University Press, Cambridge (2000b)

Chin, C.A., Barreto, A., Cremades, J.G., Adjouadi, M.: Integrated electromyogram and eye-gaze tracking cursor control system for computer users with motor disabilities. J. Rehabil. Res. Dev. **45**(1), 161–174 (2008)

Coan, J.A., Allen, J.J.B.: Frontal EEG asymmetry and the behavioural activation and inhibition systems. Psychophysiology **40**, 106–114 (2003)

Dekker, A., Champion, E.: Please biofeed the zombies: enhancing the gameplay and display of a horror game using biofeedback. Paper presented at the DiGRA, Tokyo (2007)

Ebner-Priemer, U.W., Trill, T.J.: Ambulatory assessment: an innovative and promising approach for clinical psychology. Eur. Psychol. 14(2), 109–119 (2009)

Fairclough, S.H.: Psychophysiological inference and physiological computer games. Paper presented at the ACE workshop – Brainplay'07: brain-computer interfaces and games, Salzburg (2007)

Fairclough, S.H.: BCI and physiological computing: Similarities, differences and intuitive control. Paper presented at the workshop on BCI and computer games: CHI'08 (2008)

Fairclough, S.H.: Fundamentals of physiological computing. Interact. Comput. 21, 133–145 (2009)

Fairclough, S.H., Venables, L.: Psychophysiological candidates for biocybernetic control of adaptive automation. In: de Waard, D., Brookhuis, K.A., Weikert, C.M. (eds.) Human Factors in Design, pp. 177–189. Shaker, Maastricht, The Netherlands (2004)

Fairclough, S.H., Venables, L., Tattersall, A.: The use of autonomic measures for biocybernetic adaptation. Psychophysiology 42(S1), S25 (2006)

Freeman, F.G., Mikulka, P.J., Scerbo, M.W., Scott, L.: An evaluation of an adaptive automation system using a cognitive vigilance task. Biol. Psychol. 67, 283–297 (2004)

Gevins, A., Smith, M.E., Leong, H., McEvoy, L., Whitfield, S., Du, R. et al.: Monitoring working memory load during computer-based tasks with EEG pattern recognition models. Hum. Factors 40(1), 79–91 (1998)

Gilleade, K.M., Dix, A.: Using frustration in the design of adaptive videogame. Paper presented at the advances in computer entertainment technology (2004)

Gilleade, K.M., Dix, A., Allanson, J.: Affective videogames and modes of affective gaming: Assist me, challenge me, emote me. Paper presented at the proceedings of DiGRA (2005)

Grauman, K., Betke, M., Gips, J., Bradski, G.R.: Communication via eye blinks – detection and duration analysis in real time. Paper presented at the IEEE computer society conference on Computer Vision and Pattern Recognition (CVPR'01) (2001)

Grimes, D., Tan, D.S., Hudson, S.E., Shenoy, P., Rao, R.P.N.: Feasibility and pragmatics of classifying working memory load with an electroencelograph. Paper presented at the CHI (2008)

Guger, C., Edlinger, A.G., Harkam, W., Niedermayer, I., Pfurtscheller, G.: How many people are able to operate an EEG-based brain-computer interface (BCI)? IEEE Trans. Neural Syst. Rehabil. Eng. 11(2), 145–147 (2003)

Hancock, P.A.: Teleology for technology. In: Parasuraman, R., Mouloua, M. (eds.) Automation and Human Performance, pp. 461–497. Erlbaum, Hillsdale (1996)

Hancock, P.A., Pepe, A.A., Murphy, L.L.: Hedonomics: The power of positive and pleasurable ergonomics. Ergon. Des. 13(1), 8–14 (2005)

Hancock, P.A., Szalma, J.L.: The future of neuroergonomics. Theor. Issues Ergon. Sci. 44(1–2), 238–249 (2003)

Healey, J., Picard, R.W.: Digital processing of affective signals. Paper presented at the proceedings of ICASSP, Seatle, WA (1997)

Helander, M.G., Tham, M.P.: Hedonomics – affective human factors design. Ergonomics 46(13/14), 1269–1272 (2003)

Hettinger, L.J., Branco, P., Encarnaco, L.M., Bonato, P.: Neuroadaptive technologies: applying neuroergonomics to the design of advanced interfaces. Theor. Issues Ergon. Sci. 4(1–2), 220–237 (2003)

Huang, C.N., Chen, C.H., Chung, H.Y.: Application of facial electromyography in computer mouse access for people with disabilities. Disabil. Rehabil. 28(4), 231–237 (2006)
Kapoor, A., Burleson, W., Picard, R.W.: Automatic prediction of frustration. Int. J. Hum. Comput. Stud. 65, 724–736 (2007)
Karwowski, W.: Symvatology: the science of artifact-human compatibility. Theor. Issues Ergon. Sci. 1(1), 76–91 (2000)
Kelly, K.: The technium. Available from http://www.kk.org/thetechnium/archives/2006/02/symmetrical_and.php (2006)
Klein, G., Woods, D.D., Bradshaw, J.M., Hoffman, R.R., Feltovich, P.J.: Ten challenges for making automation a "team player" in joint human-agent activity. IEEE Intell. Syst. 19(6), 91–95 (2004)
Kosmack Vaara, E., Hook, K., Tholander, J.: Mirror bodily experiences over time. Paper presented at the CHI, Boston (2009)
Kulic, D., Croft, E.: Anxiety detection during human-robot interaction. Paper presented at the conference on intelligent robots and systems, Edmonton, Canada (2005)
Kulic, D., Croft, E.: Estimating robot induced affective state using Hidden Markov Models. Paper presented at the IEEE international symposium on robot and human interactive communication (2006)
Lisetti, C.L., Nasoz, F.: Using noninvasive wearable computers to recognize human emotions from physiological signals. EURASIP J. Appl. Signal Process. 11, 1672–1687 (2004)
Lisetti, C.L., Nasoz, F., LeRouge, C., Ozyer, O., Alvarez, K.: Developing multimodal intelligent affective interfaces for tele-home health care. Int. J. Hum. Comput. Stud. 59, 245–255 (2003)
Liu, C., Rani, P., Sarkar, N.: An empirical study of machine learning techniques for affect recognition in human-robot interaction. Paper presented at the IEEE conference on intelligent robots and systems (2005)
Mandryk, R.L., Atkins, M.S.: A fuzzy physiological approach for continuously modeling emotion during interaction with play technologies. Int. J. Hum. Comput. Stud. 65, 329–347 (2007)
Mandryk, R.L., Inkpen, K.M., Calvert, T.W.: Using psychophysiological techniques to measure user experience with entertainment technologies. Behav. Inform. Technol. 25(2), 141–158 (2006)
McFetridge-Durdle, J.A., Routledge, F.S., Parry, M.J.E, Dean, C.R.T, Tucker, B.: Ambulatory impedance cardiography in hypertension: a validation study. Eur. J. Cardiovasc. Nurs. 7, 204–213 (2008)
Merrill, D., Maes, P.: Augmenting looking, pointing and reaching gestures to enhance the searching and browsing of physical objects. Paper presented at the 5th international conference on pervasive computing (Pervasive'07), Toronto (2007)
Microsoft Project Natal: http://www.xbox.com/en-US/live/projectnatal/
Milenkovic, A., Otto, C., Jovanov, E.: Wireless sensor networks for personal health monitoring: Issues and an implementation. Comput. Commun. 29, 2521–2533 (2006)
Miller, C.A.: Trust in adaptive automation: the role of etiquette in tuning trust via analogic and affective methods. Paper presented at the first international conference on augmented cognition, Las Vegas, 22nd–27th July 2005
Morris, M.E.: Technologies for heart and mind: new directions in embedded assessment. Intel. Technol. J. 11(1), 67–74 (2007)
Morris, M., Guilak, F.: Mobile heart health: project highlight. IEEE Pervasive Comput. 8(2), 57–61 (2009)
Murata, A.: Eye gaze input versus mouse: cursor control as a function of age. Int. J. Hum. Comput. Interact. 21(1), 1–14 (2006)
Naik, G.R., Kumar, D.K., Arjunan, S.P.: Multi modal gesture identification for HCI using surface EMG. Paper presented at the 12th international conference on entertainment and media in the ubiquitous era, Tampere (2008)
Norman, D.A.: The Design of Future Things. Basic Books, New York, NY (2007)

O'Donnell, R.D., Eggemeier, FT: Workload assessment methodology. In: Boff, K., Kaufman, L., Thomas, J. (eds.) Handbook of Human Perception and Performance, vol. 2, pp. 42.41–42.49. Wiley, New York, NY (1986)
Picard, R.W.: Affective Computing. MIT Press, Cambridge (1997)
Picard, R.W., Healey, J.: Affective wearables. Personal Technol. 1, 231–240 (1997)
Picard, R.W., Klein, J.: Computers that recognise and respond to user emotion: theoretical and practical implications. Interact. Comput. 14, 141–169 (2002)
Picard, R.W., Vyzas, E., Healey, J: Towards machine emotional intelligence: analysis of affective physiological state. IEEE Trans. Pattern Anal. Mach. Intell. 23(10), 1175–1191 (2001)
Pope, A.T., Bogart, E.H., Bartolome, D.S.: Biocybernetic system evaluates indices of operator engagement in automated task. Biol. Psychol. 40, 187–195 (1995)
Prinzel, L.J., Freeman, F.G., Scerbo, M.W., Mikulka, P.J., Pope, A.T.: A closed-loop system for examining psychophysiological measures for adaptive task allocation. Int. J. Aviat. Psychol. 10(4), 393–410 (2000)
Prinzel, L.J., Freeman, F.G., Scerbo, M.W., Mikulka, P.J., Pope, A.T.: Effects of a psychophysiological system for adaptive automation on performance, workload, and the event-related potential P300 component. Hum. Factors. 45(4), 601–613 (2003)
Prinzel, L.J., Parasuraman, R., Freeman, F.G., Scerbo, M.W., Mikulka, P.J., Pope, A.T.: Three experiments examining the use of electroencephalogram, event-related potentials, and heart-rate variability for real-time human-centred adaptive automation design (No. NASA/TP-2003-212442): NASA (2003)
Prinzel, L.J., Pope, A.T., Freeman, F.G.: Physiological self-regulation and adaptive automation. Int. J. Aviat. Psychol. 12(2), 179–196 (2002)
Rani, P., Sarkar, N.: Operator engagement detection for robot behaviour adaptation. Int. J. Adv. Robot Syst. 4(1), 1–12 (2007)
Rani, P., Sarkar, N., Liu, C.: Maintaining optimal challenge in computer games through real-time physiological feedback. Paper presented at the 11th human-computer interaction international, Las Vegas (2005)
Rani, P., Sims, J., Brackin, R., Sarkar, N.: Online stress detection using psychophysiological signal for implicit human-robot cooperation. Robotica 20(6), 673–686 (2002)
Rashidi, P., Cook, D.J.: Keeping the resident in the loop. Adapting the smart home to the user. IEEE Trans. Syst. Man Cybern. – Part A: Syst. Hum. 38, 949–959 (2009)
Reynolds, C., Picard, R.W.: Affective sensors, privacy and ethical contracts. Paper presented at the CHI'2005, Vienna, Austria (2005a)
Reynolds, C., Picard, R.W.: Evaluation of affective computing systems from a dimensional metaethical position. Paper presented at the First Augmented Cognition Conference, Las Vegas (2005b)
Ryan, R.M., Rigby, C.S., Przybylski, A: The motivational pull of video games: a self-determination approach. Motiv. Emot. 30, 347–363 (2006)
Saponas, T.S., Tan, D.S., Morris, D., Balakrishnan, R.: Demonstrating the feasibility of using forearm electromyography for muscle-computer interfaces. Paper presented at the CHI (2008)
Scerbo, M.W., Freeman, F.G., Mikulka, P.J.: A brain-based system for adaptive automation. Theor. Issues Ergon. Sci. 4(1–2), 200–219 (2003)
Scheirer, J., Fernandez, R., Klein, J., Picard, R.W.: Frustrating the user on purpose: a step toward building an affective computer. Interact. Comput. 14, 93–188 (2002)
Schwartz, M.S., Andrasik, F.: Biofeedback: A Practioners Guide. Guildford Press, New York, NY (2003)
Tecce, J.J., Gips, J., Olivieri, C.P., Pok, L.J., Consiglio, M.R.: Eye movement control of computer functions. Int. J. Psychophysiol. 29, 319–325 (1998)
Teller, A.: A platform for wearable physiological computing. Interact. Comput. 16, 917–937 (2004)
Wachsmuth, S., Wrede, S., Hanheide, M.: Coordinating interactive vision behaviours for cognitive assistance. Comput. Vis. Image Underst. 108, 135–149 (2007)

Wickens, C.D.: Multiple resources and performance prediction. Theor. Issues Ergon. Sci. 3, 150–177 (2002)

Wilson, G.F., Russell, C.A.: Operator functional state classification using multiple psychophysiological features in an air traffic control task. Hum. Factors 45(3), 381–389 (2003)

Wilson, G.F., Russell, C.A.: Performance enhancement in an uninhabited air vehicle task using psychophysiologically determined adaptive aiding. Hum. Factors 49(6), 1005–1018 (2007)

Wolpaw, J.R., Birbaumer, N., Heetderks, W.J., McFarland, D.J., Peckham, P.H., Schalk, G. et al.: Brain-computer interface technology: a review of the first international meeting. IEEE Trans. Rehabil. Eng. 8, 164–173 (2000)

Wolpaw, J.R., Birbaumer, N., McFarland, D.J., Pfurtscheller, G., Vaughan, T.M.: Brain-computer interfaces for communication and control. Clin. Neurophysiol. 113, 767–791 (2002)

Wolpaw, J.R., McFarland, D.J.: Control of a two-dimensional movement signal by a noninvasive brain-computer interface in humans. Proc. Natl. Acad. Sci. USA 101, 17849–17854 (2004)

Yannakakis, G.N., Hallam, J., Hautop Lund, H.: Entertainment capture through heart rate activity on physical interactive playgrounds. User Model. User-Adapt. Interact. 18, 207–243 (2007)

Zander, T.O., Jatzev, S.: Detecting affective covert user states with passive brain-computer interfaces. Paper presented at the Affective Computing and Intelligent Interaction (ACII'09) (2009)

Chapter 2
Unobtrusive Emotions Sensing in Daily Life

Martin Ouwerkerk

Abstract The measurement of human emotions in a daily life setting is reviewed in this chapter. In detail the hardware aspects of the Philips Research emotion measurement platform are described. The platform contains a wireless skin conductance wristband, a wireless chest strap ECG sensor and a wireless ear clip blood volume pulse sensor, which together with an internet tablet as hub form a personal wireless network. Two examples of applications, which have been evaluated in the form of concepts are presented.

2.1 Introduction

This book contains for a large part the proceedings of the second Probing Experience symposium, organized by Philips Research. Similarly a proceedings book was published for the first symposium (Westerink et al., 2008). One of the papers of this book describes the unobtrusive sensing of psychophysiological parameters (Ouwerkerk et al., 2008). The present paper builds on what is described there.

The work described in this paper is performed in Philips Research. The company Philips is a people centric company. One of the Philips brand slogans is "Designed around you". Another is "Easy to experience". To help substantiate these slogans a project was started to build capability on the sensing and interpretation of emotions. The brand slogans suggest two different branches in how Philips can benefit from knowhow on emotions. "Designed around You" claims that the products and services offered by Philips are tailored and personalized to the user. The design of their interface and appearance should take into account the emotional responses to stimuli provided by the product. Preferably this should work without a learning period for all persons irrespective of age, gender and cultural background. The "Easy to Experience" claim tells more about how the product is used by the customer. Prediction as to how usage affects the user often also involves knowhow on emotions.

M. Ouwerkerk (✉)
Brain, Body and Behavior Group, Philips Research, Eindhoven, The Netherlands
e-mail: martin.ouwerkerk@philips.com

J. Westerink et al. (eds.), *Sensing Emotions*, Philips Research Book Series 12,
DOI 10.1007/978-90-481-3258-4_2, © Springer Science+Business Media B.V. 2011

Philips is a health and well-being company. By "well-being" is meant general sense of fulfilment, feeling good and at ease. "Well-being" also refers to a sense of comfort, safety and security people feel in their environment – at home, at work, when shopping or on the road. Clearly it is important to know how people feel in these situations.

2.2 State-of-the-Art

The measurement of human emotions in a controlled laboratory environment is well studied for years by a large number of research groups. The Handbook of Affective sciences offers a good overview (Davidson et al., 2002). Usually validated stimuli are offered to young psychology students and their responses are carefully monitored by questionnaires, interviews, or in some cases by recording their psychophysiological parameters or facial expressions. Occasionally the psychology students are replaced by specific occupational groups, such as airline pilots, military personnel, radar operators.

The art of monitoring human behaviour in their own ecosystem is generally called ambulatory assessment. It comprises the use of field methods to assess the ongoing behaviour, physiology, experience and environmental aspects of humans or non-human primates in naturalistic or unconstrained settings. Ambulatory assessment designates an ecologically relevant assessment perspective that aims at understanding bio psychosocial processes as they naturally unfold in time and in context.

Ambulatory assessment covers a range of methodologies of real-time data capture that originate from different scientific disciplines. These methodologies include but are not limited to experience sampling, repeated-entry diary techniques, and ambulatory monitoring of physiological function, physical activity and/or movement, as well as the acquisition of ambient environmental parameters.

A society for ambulatory assessment has recently been launched. See http://www.ambulatory-assessment.org/ for contact details. A recent overview of the current techniques and challenges is published by Jochen Fahrenberg (Fahrenberg et al., 2007).

This paper refers to the use of computer-assisted methods for self-reports, behaviour records, or physiological measurements, while the participant performs normal daily activities. In recent decades, portable microcomputer systems and physiological recorders/analyzers have been developed for this purpose. In contrast to their use in medicine, these new methods have so far hardly entered the domain of psychology. In spite of the known deficiencies of retrospective self-reports, questionnaire methods are often still preferred. The most common assessment approaches are continuous monitoring, monitoring with time- and event-sampling methods, in-field psychological testing, field experimentation, interactive assessment, symptom monitoring, and self-management. Such approaches address ecological validity, context specificity, and are suitable for practical applications.

Whereas the ambulatory assessment society focuses at behavioural research, there are also research groups dedicated solely to the study of emotions. The Geneva Emotion Research Group, directed by Prof. Klaus R. Scherer, works on theoretical development and empirical research in the affective sciences, with special emphasis on emotion-constituent appraisal processes, expression of emotion and stress in voice and speech, facial expression of emotion, central and peripheral physiological reaction patterns, and subjective experience of emotional processes. Their research methods include experimental studies in both laboratory and field settings, using emotion induction and sampling of naturalistic emotions, as well as computer-simulation approaches.

Although for extending the level of scientific understanding of emotions this research is of great importance its benefits for understanding the experience of products, personalized design and the well-being of persons are bound to be limited. In a daily life setting the measurement of emotions has some extra challenges. For instance the existence of multiple simultaneous stimuli, the effects of emotions building on previously experienced emotions, social effects, and the effects of the presence of an emotion sensor on the observed emotion.

It is the objective of this Chapter to address the issues linked to sensing emotions in a daily life setting. An emotion sensing platform has been developed to serve as a research tool for this purpose (Westerink et al., 2009). The design considerations of the hardware components, as well as some first experiences will be discussed.

2.2.1 Wearable Physiology Sensor Devices

A review on the available wearable physiology sensor devices, which can be used for emotion sensing can be found in a description of recording methods in applied environments (Fahrenberg, 2000). Pioneering work on the measurement of emotions while driving a car was done at MIT by Jennifer Healey (Healey et al., 1999).

A good overview on which physiological parameters are responding to emotional stimuli can be found in (Boucsein, 2000). One of emotion related physiological parameters is the skin conductance. The group of Rosalind Picard at the Medialab of the Massachusetts Institute of Technology has worked for years on the development of a wearable skin conductance sensor. First a glove was made, called the Galvactivator, which senses skin conductivity and communicates the level changes via LEDs (Picard and Scheirer, 2001). Here the skin contacts were placed at the palm of the hand. Later the iCalm skin conductance sensing wristband was made (see Fig. 2.1), which contacted the underside of the wrist (Poh et al., 2009). This device is now planned to be productized under the name Affectiva Q.

In Germany a skin conductance sensing wristband, called the “smartband”, is offered by bodymonitor.de (http://www.bodymonitor.de/). This device records the raw sensor data of an entire day. At the end of the day the data is offloaded to a PC for analysis.

Fig. 2.1 The iCalm skin conductance sensing wristband made by MIT

A recent and complete overview on the topic of body sensor networks has been edited by Guang-Zhong Yang of Imperial College London (Yang, 2007). The emphasis is on healthcare applications. Topics like energy scavenging, context-aware sensing, multi-sensor fusion and autonomic sensing are of interest to the application described here.

Peter et al. describe a wireless sensor network for so-called "intelligent" acquisition and processing of physiological information aimed at inferring information on the mental and emotional state of a human (Peter et al., 2005). Aspects such as a being light-weight, being non-intrusive for the user and ease of use are very similar to the objectives of the Philips Emotion Sensing Platform. The glove of this platform has a skin resistance sensor as well as a temperature sensor. The heart rate is obtained from a Polar heart rate chest belt. Regretfully no recent publications showing data obtained using this system were found.

The Nexus-10 from MindMedia is a wearable device capable of sensing the galvanic skin response, ECG and respiration simultaneously for an entire day. The data is either logged onto an SD flash card or transmitted by a Bluetooth transceiver to a PC with the appropriate data management software. The dimensions of the device are 11.4 × 9.8 × 3.7 cm (l × b × d) and the weight 165 g. The sensor leads are hardwired to the device. The device is worn in a belt pouch.

2.2.2 From Physiology Sensing to Emotional State Assessment

The objective of assessing the emotional state of a person in daily life brings several challenges. Obviously, the presence of the emotion sensing system should not influence the emotions of the wearer. The system therefore is designed to be unobtrusive and as unnoticeable as possible.

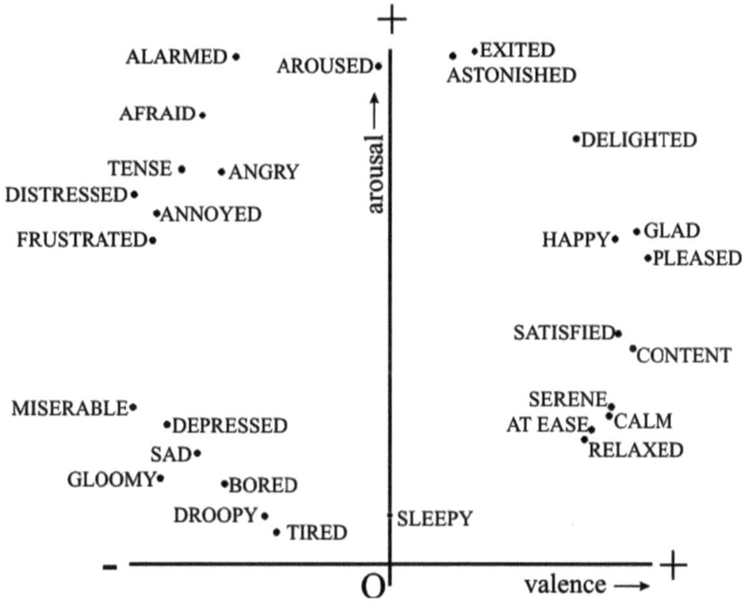

Fig. 2.2 Valence-arousal classification of emotions

Emotions usually are short lived, contrary to moods or character traits. Classification of emotions may be done in a 2 dimensional space formed by an intensity or arousal axis and a valence axis as pointed out by Russell (Russell 1989). This is shown in Fig. 2.2. An emotional status may be described by a position in this 2-dimensional space.

The most precise way to monitor relevant psychophysiological signals is to monitor hormones linked to emotions, such as the stress hormone cortisol, or DHEA (dehydroepiandrosterone) (Grillon et al., 2006). Saliva testing is currently the most used method available for this. Due to the large time lag of saliva sampling and obtaining the lab result, real-time monitoring of the hormone levels is currently not feasible in a daily life situation.

Alternatively, the monitoring of the effects of these regulatory body chemicals provides insight into a person's emotional status. For instance an emotional response to a stimulus, which leads to a higher arousal level, results in a number of psychophysiological changes, such as emotion evoked sweating, heart rhythm changes, muscle tension increases, and respiration rate changes (Boucsein, 2000). Real-time monitoring of these parameters therefore can provide information on changes in the arousal level of a person.

Obtaining accurate information on the valence of emotions is still necessary for the assessment of the emotion status. Usually in psychological tests questionnaires provide insight into this. For an unobtrusive emotion sensing system for daily life use this can not be done. Another way of obtaining information on this is the interpretation of facial expression, either by monitoring muscle activity of facial muscles or video image analysis. Both seem only partially accessible in daily life situations. Discriminating information on the valence of a person's emotions may come from data on the nature of activity a person is engaged in. Emotion sensing parameters can be differentiated as to whether they provide information or quantification of the valence aspect or the arousal aspect (see Table 2.1). In Table 2.2 the methods for obtaining information or quantification of the relevant parameters for emotions sensing are shown.

Suitable parameters for emotion sensing meant for use in daily life are emotion induced sweating, heart rate variability, voice intonation and context assessment. A more or less complete emotion sensing system therefore comprises a skin conductance sensor, a 3D accelerometer, an ECG sensor, a microphone, and a small camera worn on the chest.

The Emotion Sensing and Interpretation project that ran from 2007 until end of 2008 at Philips Research aimed at a miniature wireless sensor system capable of sensing emotions during an entire day without maintenance. The devices ere meant to be unobtrusive, i.e. so small and lightweight as to not to be noticeable by the wearer. The shape of the device package was to be optimized for the position on the body where it is to be worn.

Table 2.1 Emotion related parameters that can be linked (marked with +) to arousal and/or valence

Emotion-related effect	Arousal	Valence
Emotion induced sweating	+	
Breathing rhythm variations	+	+
Heart rate variability	+	+
Blood pressure	+	
Core body temperature	+	
Heart rate	+	
Facial expression		+
Facial muscle activity such as jaw clenching	+	
Voice intonation	+	+
Questionnaire	+	+

Table 2.2 Relevant parameters for emotions sensing and the method for obtaining them

Parameter	Method
Emotion induced sweating	Skin conductance
Breathing rhythm variations	Respiration sensor
Heart rate variability (HRV)	Electrocardiography
Blood pressure, HRV	Photoplethysmography
Brain activity	Electroencephalography
Muscle tension, jaw clenching	Electromyography
Core body temperature	Temperature sensor
Facial expression	Camera and computer
Voice intonation	Microphone and computer
Context assessment	Accelerometer, camera, microphone and computer

This emotion sensing system will be used on people in daily life settings to collect data. In a separate set of experiments test persons have been subjected to a series of predetermined emotional stimuli and their emotional response were recorded to serve as a yardstick for the data interpretation with this system.

2.2.3 Design of Wearable Emotion Sensors

The design considerations for unobtrusive sensing of psychophysiological parameters are described in a chapter of a book on probing experience (Ouwerkerk et al., 2008), Oliver Amft of ETH Zurich has published a paper on the design of miniature wearable systems (Amft et al., 2004), such as a system integrated in a shirt button. He pioneered in exploring the system architecture of this and similar wearable sensor systems. The emotions sensing system described here needs to comply with the following requirements: it needs to be unobtrusive, robust against motion artefacts and outdoor and daily life use, and maintenance friendly. Gemperle determined the optimal areas on the body for unobtrusive devices (http://www.ices/cmu.edu/design/wearability/). The areas found to be the most unobtrusive for wearable objects are: collar area, rear of the upper arm, forearm, rear, side and front of the ribcage, waist and hips, thigh, shin, and the top of the foot Important design aspects for the design of wearable devices as identified by Gemperle are collected in Table 2.3.

When considering the above mentioned attention points, in order to maximize unobtrusiveness the devices need to be lightweight, shaped to the body, small and colored as to fit with its surroundings. The surface finish needs to be pleasant to the touch. Irritations of the skin due to prolonged use need to be avoided.

The devices need to be designed for daily use over a period of several years, requiring the package to be able to cope with mechanical impact and moisture. The electronics need to be detachable from the textile and skin contact parts, to allow cleaning. The sensing of emotion related parameters needs to be stable to

Table 2.3 Relevant parameters for wearable devices

Parameter	Description
Placement	Where on the body it should go
Form language	Defining the shape
Human movement	Consider the dynamic structure
Proxemics	Human perception of space
Sizing	For body size diversity
Attachment	Fixing forms to the body
Containment	Consider what is in the form
Weight	As it is spread across the human body
Accessibility	Physical access to the forms
Sensory interaction	For passive or active input
Thermal	Issues of heat next to the body
Aesthetics	Perceptual appropriateness
Long-term use	Effects on body and mind

motion of the wearer. The wireless link between the system and the devices needs to be reliable. The electronics power budget and battery size specification should enable entire day use without maintenance like for instance battery recharging. The electronics design should optimize the ease of use in terms of wireless software maintenance, and data download from flash during the night in a special device cradle.

2.3 Philips Emotion Sensing Platform

In emotion literature, there is consensus on the relationship between skin conductance (SC) and emotional arousal, see for instance the work by Lang (Lang et al., 1993). Although less clear, there also is evidence of a relationship between the heart rate variability derived from electrocardiogram (ECG) and emotional valence (Frazier et al., 2004 and Sakuragi, 2002). Therefore as the most promising psychophysiological parameters, SC and ECG measurement were implemented on the platform. The platform consists of a SC wrist band, an ECG chest belt and a Nokia N810 internet tablet, to be worn in a dedicated holder clipped to participants belt, serving as a hub (see Figs. 2.3 and 2.12).

The electronics modules of the SC and ECG sensor nodes are identical. Depending on the cradle they are put into, the SC or ECG measurement is activated on the node (see Fig. 2.4b). The capability to remove the electronics from the system parts that are in contact with the body facilitates cleaning or replacement of these parts, as well as battery recharging.

Twisting the electronics module in the cradle switches the device on, or off. The electronics module is shown in Fig. 2.4a.

The electronics module is designed such that it can be glued shut, to prevent exposure of the sensitive electronics. The thirteen spring contacts provide all connections needed for software upload, battery recharging and sensor skin contact pads. One pin is reserved for an auxiliary input. This can in a future extension be used for an external temperature sensor, or a respiration sensor.

These contacts also include some general purpose digital I/O as well as UART transmit and receive. This allows the device to control actuators, or to high speed transfer data from flash or RAM to other devices.

The contacts extend several millimetres from the package, facilitating the clipping of leads for testing, battery charging, as well as easy detachment from the cradle (1 N ejection force per spring contact).

The button shaped module has been kept as small (35 mm diameter) and lightweight (13.5 g) as possible within limitations of the requirements. The 160 mAh 3.7 V Lithium polymer rechargeable battery (type GMB401730 of Guangzhou Markyn Battery Co, Ltd.) enables a full day of use. The Texas

Instruments CC2420 transceiver is IEEE 802.15.4 compliant. The MAC (Media Access Control) runs on the NxH1200/CoolfluxDSP (NXP).

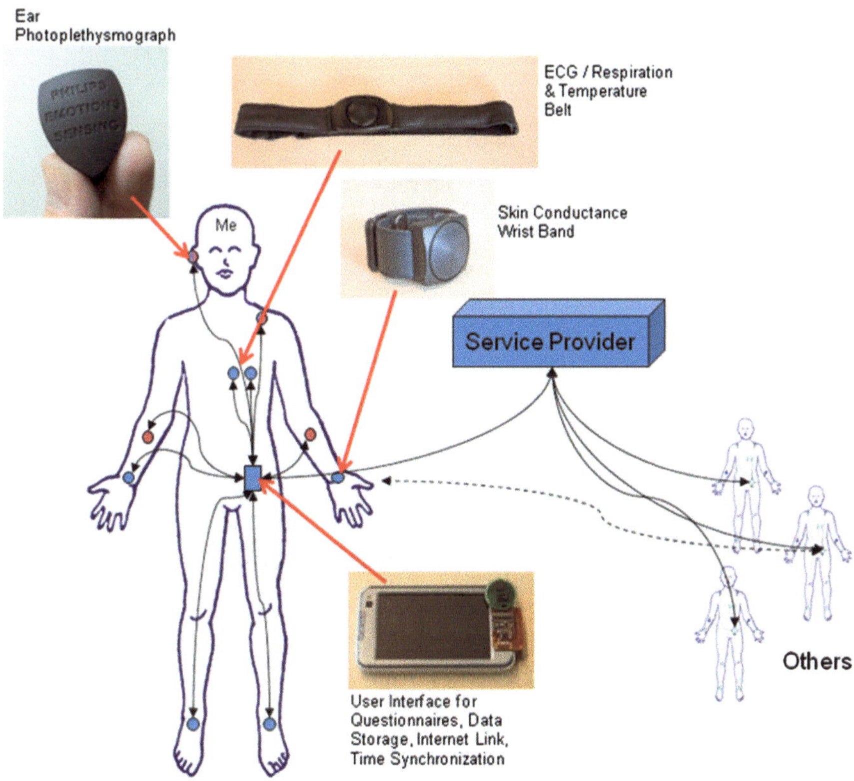

Fig. 2.3 Emotion measurement platform of Philips research

This DSP is C-programmable, has 448 kB RAM, and runs at 20 MHz. The real-time operating system FreeRTOS is used. These capabilities allow signal processing and interpretation to be done embedded in the electronics module. The combination of local processing plus the wireless transmission of an emotion status, or of emotion changes requires significantly less power compared to wireless streaming of sensor data (Ouwerkerk et al., 2006). Two megabytes of flash memory are available for data and programming code storage(serial flash M25P16 Numonyx), allowing a full day of skin conductance and 3D accelerometer data to be stored at 2 samples per second. The sensor electronics are positioned on a separate printed circuit board, allowing the DSP/transceiver/flash PCB to be re-used for other sensor devices, such as a posture sensor (Ouwerkerk et al., 2006). A stand-alone linear Li-Ion battery charger including a micro power comparator is built in. This enables fast and safe battery charging from a USB socket of a PC.

The sensor board contains a 3D-accelerometer with a sensitivity of about 12 mm/s^2 (Kionix KXSD9), and the electronics for skin conductance and ECG sensing. Additionally an onboard temperature sensor (National Semiconductor LM19), and a battery voltage sensor are present. The skin conductance is sensed

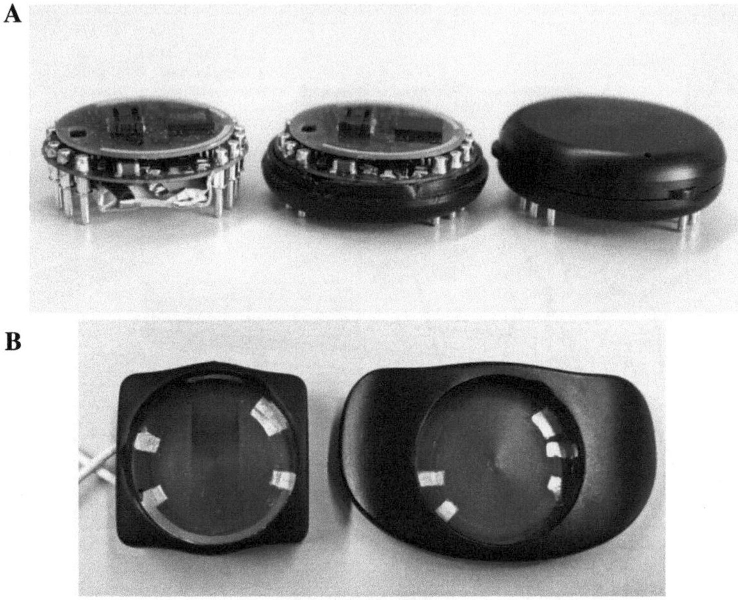

Fig. 2.4 Electronics module for skin conductance and ECG sensing (**a**), and cradles (**b**) for Skin conductance sensor (*left picture*) and ECG sensor (*right picture*) showing contacts for sensor type and device activation

by a DC current, using a reference voltage source, and 3 switchable reference resistances to allow the measurement of a wide range of skin resistances current, using a reference voltage source, and 3 switchable reference resistances to allow the measurement of a wide range of skin resistances

The ECG sensor electronics contains an amplification stage (100× gain), and a 2nd order Butterworth low-pass filter with a cut off frequency at 80 Hz. Appropriate safety measures are taken.

The ECG/SC module is a packaged wireless sensor device with built-in antenna. The efficiency of the antenna has been measured and modelled for body network applications (Alomainy et al., 2007).

The device is tested to be IEC60101-1 compliant.

The sensed battery voltage, the skin conductance signal, and the ECG signal are digitized by a 12-bit analog to digital convertor. In the choice of components a trade of between signal quality and low power usage was made, to ensure full day of operation on the battery.

2.3.1 Skin Conductance Sensor

The skin contacts were positioned at the underside (volar) of the wrist, because at that position the skin does not have hairs. The skin conductance level at the volar side of the wrist is 0.36 times the value obtained at the standard location at the fingertips (Table 2.1 of Edelberg, 1967). The measurement of the skin conductance at the wrist is therefore possible. Upon comparing the signals at the wrist and the fingertips it was found that the smaller emotion related responses could be detected at the fingertips, but not at the volar side of the wrist. The more pronounced effects were observable at both locations. The skin conductance sensor consists of two parts: the detachable electronics module (see Fig. 2.4), and the wristband. The development of the current embodiment of the wristband went through a number of intermediate steps. In Fig. 2.5 these are shown in one picture. Version A was the first attempt to make a skin conductance wristband. The strap was not detachable, and could not accommodate all wrist sizes with good comfort. Version B had a split metal spring strap. This could indeed accommodate most wrist sizes, but the connection to the wrist was not tight enough, which gave motion artefacts in the skin conductance signal. Version C was the final version of the wristband. A Velcro tightening strap with a stretchable strip was added. In Fig. 2.6 this version is shown in detail. Two circular metal skin contacts (11 mm diameter and 3 mm apart) placed in the wrist band served to measure the participants skin conductivity. For most persons this choice of skin contact gave measurable results. For a significant part of the test participants the skin conductance was found to be very low. This can be attributed to the electronic to ionic interface where the metal part touches the skin. Since a DC current at a potential of maximal 1.2 V was used the polarization potential of this interface may be too high for some persons to achieve a measurable current. This is an issue, that remains to be solved, when a stable skin contact is needed, which functions properly for all participants under all climate circumstances.

Version E is a wired version for use with a LabVIEW data acquisition card. It uses conductive plastic electrodes as skin contact. Version F is a fingertip contact strap version of the wireless skin conductance sensor. These versions were made to study the skin conductance signal at other body parts than the underside of the wrist.

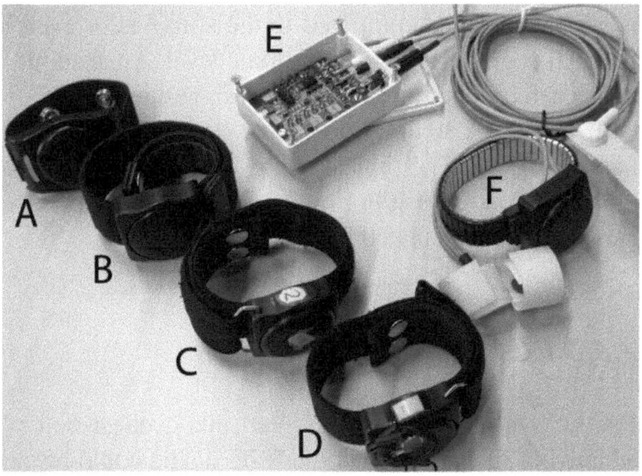

Fig. 2.5 Various embodiments of the skin conductance sensor, showing the steps leading to the final unobtrusive version. See text for a full description

Fig. 2.6 Final version of skin conductance wristband, showing tightening strap and the two metal skin contacts

For version C, the comfort of the wearers was determined in a separate study, and is reported to be comparable to wearing a watch (Westerink et al., 2009).

As already mentioned in the previous section, the skin conductance signal is sampled at the nodes at 2 Hz. The data samples, accompanied by a timestamp, can be used in a number of manners. For instance, they can be stored in flash memory. At the end of a day the collected data could be offloaded to a PC for analysis. Alternatively, the data could be streamed to the Nokia N810 hub (see Section 2.3.4) for real-time arousal event detection. Thirdly the data could be analyzed by the on-board DSP for real-time arousal detection. The version D of Fig. 2.5 has a buzzer built in, which discretely signals emotional events to the wearer. When needed, an intermittent transmission from the Nokia N810 hub can resynchronize the sensor clocks to prevent drift.
The data processing algorithms are reported in detail by Westerink et al. (2009).

Basically two approaches are possible: the first measures the tonic level, that is, the basic (averaged) level of skin conductance. The second approach considers the deviations of the signal on top of the tonic level, that is, it considers individual skin conductance responses (SCRs). These SCRs are responses caused by stimuli perceived by the wearer. They are characterized by a steep increase (of which the onset is slightly delayed compared to the stimulus or event), reaching a maximum after which it degrades slowly to the tonic level. An example of a typical skin conductance and accelerometer measurement is shown in Fig. 2.7.

In Fig. 2.7 a 30 min trace of the skin conductance trace as measured by the wrist band is shown. The typical shape of the electro dermal response is well visible in the trace. Also shown in Fig. 2.7 are traces of the 3D accelerometer output, and the activity level, calculated from this data.

2.3.2 ECG Sensor

For ECG a standard Polar WearLink® chest belt with cloth skin contacts was adapted in such a way that the ECG cradle could be easily connected (Fig. 2.8).

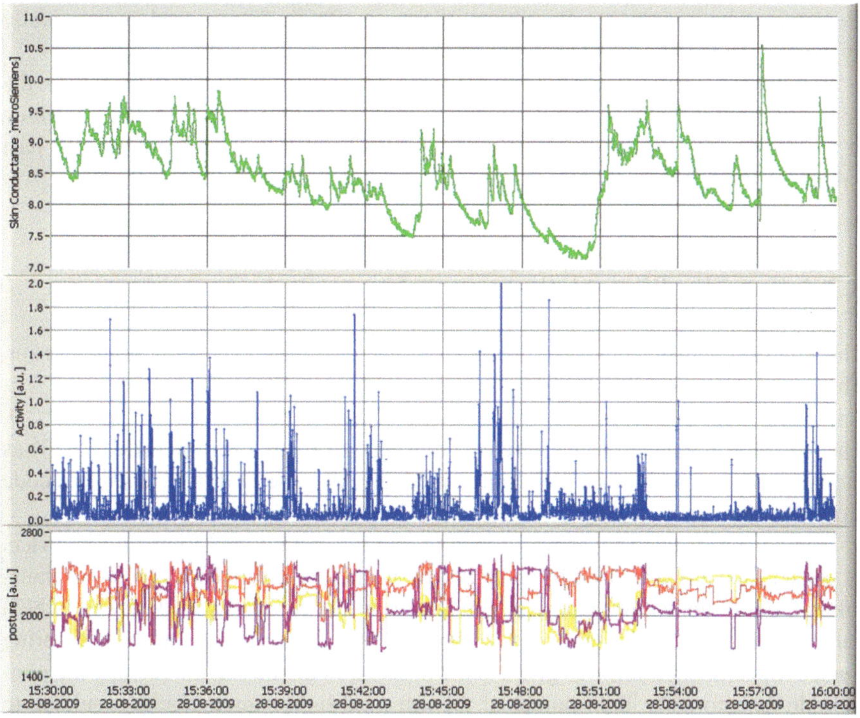

Fig. 2.7 Example of a 30 min skin conductance and 3D-accelerometer measurement obtained by the wrist sensor. The upper trace shows the skin conductance. The middle trace shows the activity level in arbitrary units, and the lower trace shows the raw output of the 3D accelerometer

The data processing algorithms for the ECG sensor are reported in detail by Westerink et al. (2009). The heart rate is determined from the ECG measured at 100 Hz. Embedded computing is used to determine the exact time of the maximum of the R-peak (tallest peak of the well known PQRST features in the ECG trace), which is interpreted as the heart beat moment.

The beat moments, along with the peak quality indicators (peak height and noise level) of detected peaks are sent to the receiver. At the receiver side, a further analysis of the detected beats is done. First, the inter-beat interval (IBI) is derived from

Fig. 2.8 Prototype ECG sensor with Polar WearLink® chest belt

the detected beat moments. Possible outliers are effectively removed automatically by examining if IBI exceed an adaptive upper or lower threshold.

Finally, the heart rate variability (HRV) is determined from the IBI signal as the power in the standard low-frequency band, ranging from 0.04 to 0.15 Hz (Brownley et al., 2000). This value is normalized using percentiles in the histogram of past values for the HRV.

2.3.3 Ear Clip Photoplethysmograph

The emotion sensing platform contains an ear clip photoplethysmograph sensor. This sensor measures the blood volume pulse of the ear lobe. A picture of the sensor, along with a picture showing how the sensor is worn is shown in Fig. 2.9. The sensor contains the same wireless transceiver (Texas Instruments CC2420) as the ECG/Skin conductance electronics module. Also the DSP and flash data/code storage is the same.

The design of this sensor was aimed at making a small and lightweight device. A folded flex foil was used to carry all the necessary electronics. To allow transmissive photoplethysmography an attachment to the flex foil was designed, which can be folded around the earlobe. Attachment of the device to the earlobe was spring loaded. Special attention was given to the prevention of stray light reaching the photodiode detector. The light output of the infra red light source can be regulated to use the best trade off between low power and good signal to noise ratio. Embedded software protocols take care of this. The flex foil has a second position for the photodiode for use in a reflective photoplethysmograph sensor (see Fig. 2.10).

In Fig. 2.11 is shown how the flex foil folds inside the package. Attention was given to the position of the antenna. In this position the antenna was on a flat part of

Fig. 2.9 Ear clip photoplethysmograph

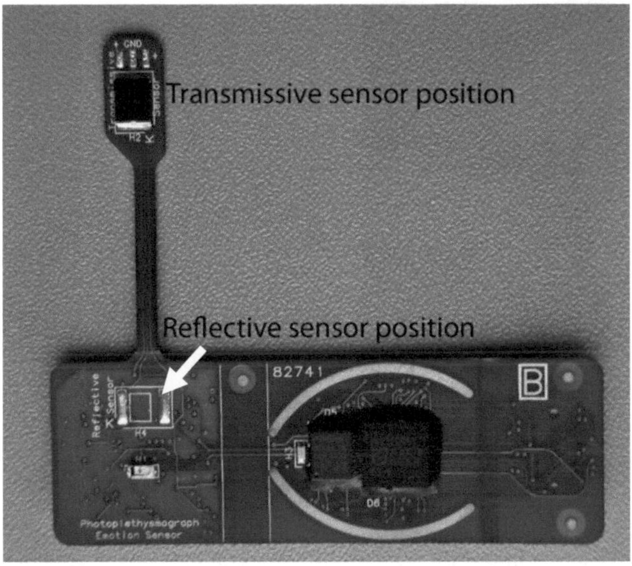

Fig. 2.10 Flex foil photoplethysmograph showing transmissive sensor diode attachment, and printed antenna. The part of the flex foil containing the connector can be cut off when software updates are no longer foreseen to save weight

the foil the most removed from the head and the metal battery. The flex foil is fitted with a 3D-accelerometer with a sensitivity of about 12 mm/s^2 (Kionix KXSD9). The part of the flex foil containing the connector can be cut off when software updates are no longer foreseen to save weight.

2.3.4 Body Network Hub

The coordinator/hub of the wireless body network is a Nokia N810 internet tablet. It is the hub of a star configured wireless personal area network. The advantage of

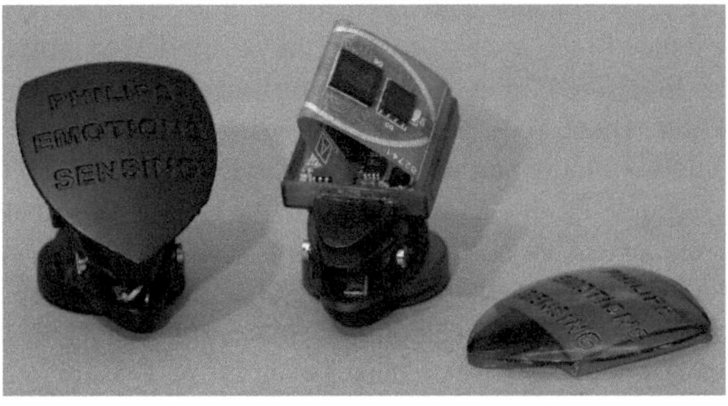

Fig. 2.11 Electronics flex foil for the ear photoplethysmograph and folding method in package

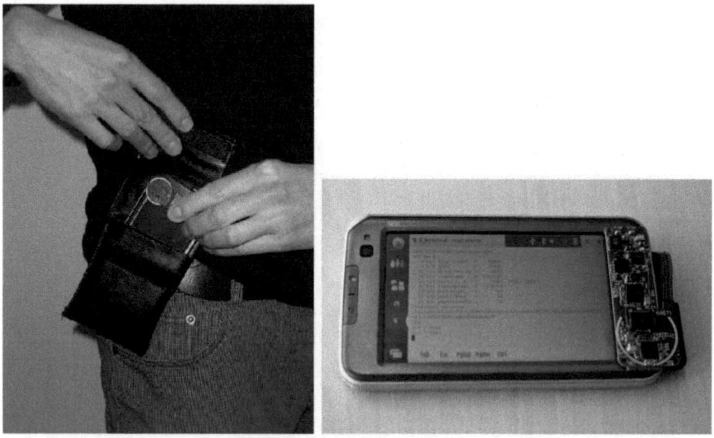

Fig. 2.12 Nokia N810 internet tablet

using this device as a hub is the availability of a graphical and a keyboard interface. This is used to offer a questionnaire to the participant at moments when a stress event is detected. Figure 2.12 shows the Nokia N810 internet tablet with a custom made add- on printed circuit board (PCB) for a IEEE 802.15.4 enabled transceiver (Texas Instruments CC2420). This board is powered via the USB connector. A special USB host-mode program is needed to make this work.

Using the same PCB a personal computer (PC) can be used as a wireless network host for these sensors. The PCB uses a FTDI USB controller (FT232RQ), which creates a virtual serial port in the PC. A middleware program available for both PC and Nokia N810 called RawDataServer creates a socket interface from which application software can use the sensor data. Application software has been written for both the MatLab and LabVIEW environment.

2.3.5 Context Determination from Accelerometer Data

As mentioned in the Sections 2.3.2 and 2.3.4 all sensor devices are fitted with a 3D-accelerometer with a sensitivity of about 12 mm/s2 (Kionix KXSD9). The orientation of the body part where the sensor device is attached can thus to some extent be monitored, by measuring the tilt of the devices using the gravitational acceleration. These body parts are the wrist for the skin conductance sensor, the plexus solaris (middle part of torso) for the ECG sensor, and the ear for the photoplethysmograph. From thus obtained data the type of activity of a person can be assessed to some extent. Robust discrimination between sitting, standing, lying down, walking and running is possible. The type of activity of a person can offer additional data in the assessment of an emotional status. This is described in detail in another chapter of this book (Bonomi, Chapter 3).

2.4 Application Examples

The emotion sensing platform is in its present form a research tool. It facilitates the monitoring of some key psychophysiological parameters related to emotions and moods. Already in two cases the emotion sensors have been built in devices, used in concept evaluation trials: the Rationalizer concept and the RelaxTV concept. These applications are described below.

2.4.1 The Rationalizer Concept

Upon a request from the ABN-AMRO bank, the so-called Rationalizer concept was developed by Philips Design (Djajadiningrat et al., 2009). It comprises a skin conductance device, based on the Philips Research emotion sensing platform wristband, offering stress level feedback to stock traders, using a LED bowl as feedback means (See Fig. 2.13).

The skin conductance sensing wristband is shown in detail in Fig. 2.14. A flex foil version of the Philips Research skin conductance wristband prototype was used along with its stress event detection software.

Fig. 2.13 Philips/ABN AMRO Rationalizer skin conductance sensor for stock dealers concept

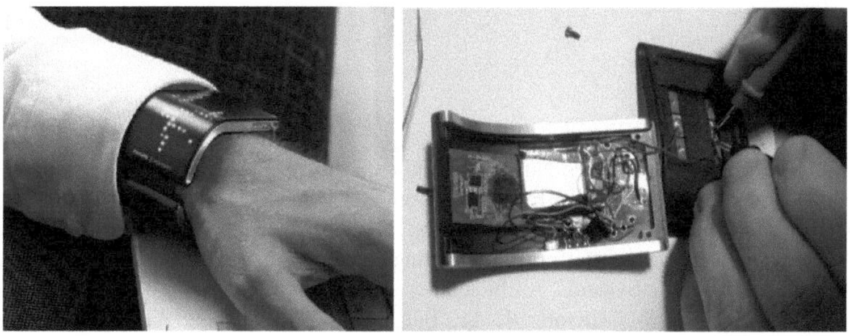

Fig. 2.14 Philips Design/ABN Amro rationalizer stress sensor for stock traders. The flex foil version of the skin conductance sensor is visible in the right picture

Stress level feedback to the wearer was partly done by LED patterns on the topside of the wristband, which contained a separate transceiver, and partly by a bowl with hundreds of programmable LEDs.

2.4.2 The RelaxTV Concept

A concept study for future television applications was done with a special hand-held sensor pebble. The so-called RelaxTV uses this photoplethysmograph sensor to obtain blood volume pulse data allowing the relaxation of a person to be monitored. Biofeedback techniques were developed to use breathing guidance for deep relaxation of a television viewer (Zoetekouw et al., 2010, RelaxTV, private communication). In Fig. 2.15 is shown how this concept looks.

Fig. 2.15 Relax TV application using photoplethysmograph blood volume sensor for heart rate sensing

Acknowledgements The author wants to express his gratitude to all persons who have contributed to the work described in this chapter, and especially to Joyce Westerink, Martijn Krans, Jack van den Eerenbeemd, Gert-Jan de Vries, Stijn de Waele, and Peter Sels.

References

Alomainy, A., Hao, Y., Pasveer, F.: Numerical and experimental evaluation of a compact sensor antenna for healthcare devices. IEEE Trans. Biomed. Circ. Syst. **1**(4), 242–249 (2007)

Amft, O., Lauffer, M., Ossevoort, S., Macaluso, F., Lukowicz, P., Troster, G.: Design of the QBIC wearable computing platform. Proceedings of the 15th IEEE International Conference on Application-Specific Systems, Architectures and Processors, pp. 398–410 (2004)

Boucsein, W., Backs, R.W.: Chapter 1. In: Backs, R.W., Boucsein, W. (eds.) Engineering Psychphysiology: Issues and Applications, pp. 3–30. Lawrence Erlbaum Associates, Hillsdale, IN (2000)

Brownley, K., Hurwitz, B., Schneiderman, N.: Cardiovascular psychophysiology. In: Cacioppo, J., Tassinary, L., Berntson, G. (eds.) Handbook of Psychophysiology, pp. 224–264. Cambridge University Press, New York, NY (2000)

Davidson, R.J., Scherer, K.R., Goldsmith, H.H. (eds.): Handbook of Affective Sciences. Oxford University Press, New York and Oxford (2002)

Djajadiningrat, T., Geurts, L., Munniksma, P.-R., Christiaansen, G., de Bont, J.: Rationalizer: an emotion mirror for online traders. Proceedings of the 5th International Workshop on Design and Semantics of Form and Movement, pp. 39–48 (2009)

Edelberg, R.: Chapter 1: electrical properties of the skin. In: Brown, C.C. (ed.) Methods in Psychophysiology, pp. 1–53. Williams & Wilkins, Baltimore, MD (1967)

Fahrenberg, J., Myrtek, M., Pawlik, K., Perrez, M.: Ambulatory assessment –monitoring behaviour in daily life settings: a behavioral-scientific challenge for psychology. Eur. J. Psychol. Assess. **23**(4), 206–213 (2007)

Fahrenberg, J., Wientjens, C.J.E.: Chapter 5. In: Backs, R.W., Boucsein, W. (eds.) Engineering Psychphysiology: Issues and Applications, pp. 111–136. Lawrence Erlbaum Associates, Hillsdale, IN (2000)

Frazier, T.W., Strauss, M.E., Steinhauer, S.R.: Respiratory sinus arrhythmia as an index of emotional response in young adults. Psychophysiology **41**, 75–83 (2004)

Grillon, C., Pine, D.S., Baas, J.M., Lawley, M., Ellis, V., Charney, D.S.: Cortisol and DHEA-S are associated with startle potentiation during aversive conditioning in humans. Psychopharmacology **186**(3), 434–441 (2006)

Healey, J., Seger, J., Picard, R.W.: Quantifying driver stress: developing a system for collecting and processing bio-metric signals in natural situations. Proceedings of the Rocky-Mountain Bio-engineering Symposium, 16–18 Apr 1999

Lang, P., Greenwald, M., Bradley, M., Hamm, O.: Looking at pictures: affective, facial, visceral, and behavioral reactions. Psychophysiology **30**, 261–273 (1993)

Ouwerkerk, M., Langereis, G., Pasveer, W.F.: Unobtrusive sensing of psychophysiological parameters, probing experience. In: Westerink, J.H.D.M., Ouwerkerk, M., Overbeek, Th.J.M., Pasveer, W.F., de Ruyter, B. (eds.) Philips Research Book Series, vol. 8, pp. 161–191. Springer, The Netherlands (2008)

Ouwerkerk, M., Pasveer, W.F., Engin, N.: SAND: a modular application development platform for miniature wireless sensors. Proceedings of BSN 2006 International Workshop on Wearable and Implantable Body Sensor Networks, pp. 166–170 (2006)

Peter, C., Ebert, E., Beikirch, H.: Wireless sensor network for "intelligent" acqui-sition and processing of physiological information. Proceedings of the 10th IEEE International Conference on Emerging Technologies and Factory Automation: ETFA, pp. 657–660 (2005)

Picard, R.W., Scheirer, J.: The galvactivator: a glove that senses and communicates skin conductivity. Proceedings from the 9th International Conference on Human-Computer Interaction, August 2001, New Orleans, LA, pp. 1538–1542 (2001)

Poh, M.Z., Swenson, N.C., Picard, R.W.: "Comfortable Sensor Wristband for Ambulatory Assessment of Electrodermal Activity," 1st Biennial Conference of the Society for Ambulatory Assessment. Greifswald, Germany (2009)

Russell, J.A.: Measures of emotion. In: Plutchnik, R., Kellerman, H. (eds.) The Measurement of Emotions, vol. 4, p. 83. Academic, New York, NY (1989)

Sakuragi, S., Sugiyama, Y.,, Takeuchi, K.: Effects of laughing and weeping on mood and heart rate variability. J. Physiol. Anthropol. Appl. Hum. Sci. **21**, 159–165 (2002)

Westerink, J.H.D.M., Ouwerkerk, M., Overbeek, Th.J.M., Pasveer, W.F., de Ruyter, B. (eds.): Probing Experience, Philips Research Book Series, vol. 8. Springer, The Netherlands (2008)

Westerink, J.H.D.M., Ouwerkerk, M., de Vries, G.-J., de Waele, S., van den Eerenbeemd, J.M.A., van Boven, M.: Emotion measurement platform for daily life situations. Paper 120 of the Proceedings of the International Conference on Affective Computing and Intelligent Interaction, Amsterdam (2009)

Yang, G.Z. (ed.): Body Sensor Networks. Springer, London (2007)

Chapter 3
Physical Activity Recognition Using a Wearable Accelerometer

Alberto G. Bonomi

Abstract Physical activity recognition represents a new frontier of improvement for context-aware applications, and for several other applications related to public health. Activity recognition requires the monitoring of physical activity in unconfined environments, using automatic systems supporting prolonged observation periods, and providing minimal discomfort to the user. Accelerometers reasonably satisfy these requirements and have therefore often been employed to identify physical activity types. This chapter will describe how the different applications of activity recognition would influence the choice of the on-body placement and the number of accelerometers. After that it will be analyzed which sampling frequency is necessary to record an acceleration signal for the purpose of activity pattern recognition, and which is the optimal strategy to segment the recorded signal to improve the recognition performance in daily life. In conclusion, it will be discussed how the user friendliness of accelerometers is influenced by the classification algorithm and by the data processing required for activity recognition.

3.1 Introduction

Activity recognition represents a new wave of interest in context-aware applications. Context awareness is defined as the ability of certain systems to adapt their behavior based on the users' activity, the users' social situation and location, which are automatically detected. A context-aware system is able to provide the user with relevant information, trigger other applications, or interact with the user in relation to future events. Activity recognition represents, therefore, an important component of the system oriented at detecting the situation that involves the user, based on which an application should respond with a specific behavior.

Activity recognition has recently become important in the area of activity monitoring for public health purposes. Indeed, the progressive decline in the physical activity level due to the adoption of sedentary lifestyles has been associated with the increasing incidence of obesity, diabetes, and cardiovascular diseases (Ekelund

A.G. Bonomi (✉)
Medical Signal Processing Group, Philips Research, Eindhoven, The Netherlands
e-mail: alberto.bonomi@philips.com

J. Westerink et al. (eds.), *Sensing Emotions*, Philips Research Book Series 12,
DOI 10.1007/978-90-481-3258-4_3, © Springer Science+Business Media B.V. 2011

et al., 2008; Hu et al., 2003). Therefore, physical activity has frequently been recommended to improve health and reduce risks for chronic diseases (Haskell et al., 2007). Consequently, an accurate and objective assessment of physical activity in daily life is necessary to determine the effectiveness of interventions aimed at increasing physical activity, and to define the amount of physical activity necessary to obtain specific health benefits (Bonomi et al., 2009a). Furthermore, the assessment of physical activity can be used as a motivational tool to guide individuals towards the adoption of a more active lifestyle. In light of this, activity recognition represents a promising tool for improving the accuracy of systems aimed at understanding and measuring the individual's physical activity behavior.

In addition, with an ageing population the incidence of falls is increasing. Some studies have shown that the earlier a fall is reported the lower the rate of morbidity and mortality it may cause (Gurley et al., 1996; Wild et al., 1981). Clearly, the use of systems which can accurately detect a fall and automatically activate an alarm or call for help could be of major benefit. A fall is not an intentional movement. However, within the context of activity recognition, it can be considered a specific form of activity. Therefore, the analytical techniques used in activity classification are also applicable to fall detection systems.

The aforementioned applications of activity recognition require the monitoring of physical activity in daily life and in unconfined environments, using automatic systems supporting prolonged observation periods, and providing minimal discomfort to the user. Activity monitors based on acceleration sensors, also called accelerometers, reasonably satisfy these requirements and so have had frequent use monitoring physical activity and activity energy expenditure (Brage et al., 2004; Crouter et al., 2006; Harris et al., 2009; Melanson and Freedson, 1996; Plasqui and Westerterp, 2007), primarily in medical research. Recently, accelerometers have also been employed to record the acceleration of the body for the extraction of information used to develop classification algorithms for identifying types of physical activity. In this chapter, some of the main characteristics of accelerometers systems will be presented in relation to their application for physical activity recognition in daily life.

3.2 Placement and Number of Accelerometers

The position worn and the number of accelerometers used to monitor physical activity significantly depend on the purpose of activity recognition. Indeed, depending on the application, the measuring system might need to detect specific activities or movements. For instance, the development of a monitoring system specific for walking could require the detection of lower limb movements. The development of systems able to discriminate body postures might require the detection of position and movements of torso and thigh, while the development of systems able to monitor the behavior of personal computer users might focus on the collection of upper limbs movements (Fig. 3.1). Generally, systems based on multiple accelerometers

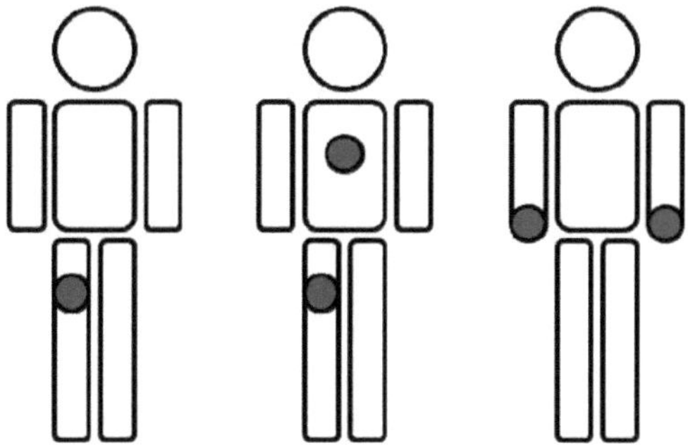

Fig. 3.1 Placement of accelerometers for specific applications of activity recognition. Monitoring of characteristics of gait and walking activities (*left*); monitoring of postures and postural changes (*middle*); monitoring of activity for personal computer users (*right*)

are able to detect a broader range of activity types as compared to systems based on a single accelerometer. The signal measured using a single accelerometer could be used to determine the engagement in more generic classes of activities, such as walking, running, sedentary occupation, or sports, while multiple-accelerometers systems could identify physical activity in a more refined way by recognizing sub-activities such as walking on a flat surface, walking upstairs, walking downstairs, Nordic walking, running, jogging, sitting, standing, or rowing, due to the fact that by using several acceleration sensors, a bigger amount of information on the body acceleration can be collected. However, the more accelerometers are necessary to record physical activity, the higher the interference of the measuring system with the spontaneous behavior of the user. Therefore, the development of simple, small and light-weight systems to monitor physical activity in daily life should be recommended, particularly for the enhanced wearability.

A network of accelerometer sensors has been used to identify a broad range of activities by analyzing the acceleration signal using artificial neural network algorithms (Zhang et al., 2003), decision trees algorithms (Bao and Intille, 2004) or thresholds-based algorithms (Veltink et al., 1996). More recently, activity recognition has been proposed by analyzing the acceleration of the body as collected using a single accelerometer (Boissy et al., 2007; Bonomi et al., 2009b; Ermes et al., 2008a; Karantonis et al., 2006; Mathie et al., 2003; Pober et al., 2006;). Accurate classification performances have been observed in identifying walking, running, and cycling (Bonomi et al., 2009b). However, the simplification of the measurement system resulted in a decrease in the ability of correctly identifying certain types of activities, such as sitting and standing, as compared with the performance of multiple accelerometer systems.

3.3 Optimal Sampling Frequency

The recognition of type of physical activity requires the collection of detailed information of the acceleration pattern of the body. This means that accelerometers employed for activity recognition usually sample the body acceleration with an adequately high frequency. Some studies that investigated the body center of mass acceleration signal during walking showed that 95% of the signal could be determined by harmonics within 10 Hz (Antonsson and Mann, 1985). However, there is not a general guideline as to which sampling frequency should be used to monitor physical activity for identifying a broad range of activity types. In literature, accurate classification of physical activity has often been achieved using accelerometers that sampled the acceleration signal with a frequency of 20, 32 or 50 Hz (Bao and Intille, 2004; Bonomi et al., 2009b; Ermes et al., 2008b; Ermes et al., 2008a). Consequently, this accelerometer was characterized by relatively high power consumption and a battery life that usually did not exceed a few days. Low power consumption has been the major design constraint for wearable accelerometers (Ermes et al., 2008a). Improvements in battery life can be achieved by developing less power consuming accelerometers as well as activity recognition algorithms that process a reduced amount of data and operate with a reduced sampling frequency. To address this problem, in our research laboratory, a decision tree algorithm was developed to identify a series of sedentary, locomotive, housework and sports related activities by measuring the acceleration of the body using a single tri-axial accelerometer placed on the lower back. The aim was to investigate whether the use of a low sampling frequency could enable the development of classification models with high accuracy. For this purpose, fifteen healthy young adults (10 males and 5 females, age: 32.8 ± 8.2 y; body mass: 74.2 ± 13.4 kg; height: 1.78 ± 0.08 m; body mass index: 23.2 ± 3.4 kg/m^2) were recruited. Physical activity of these volunteers was measured using an activity monitor equipped with a tri-axial piezo-capacitive sensor (KXP84, Kionix, Ithaca, NY). The device recorded the raw acceleration signal with a sampling frequency of 20 Hz, and the battery life was 36 h. These test subjects completed a protocol consisting of a series of 14 standardized activities (Fig. 3.2).

The acceleration signal, recorded during the experimental trial at 20 Hz, was sub-sampled at 10, 5, 2.5, and 1.25 Hz by decimation after proper anti-aliasing filtering. From these 5 derived signals, the data segment corresponding to each activity task was isolated. The isolated data was labeled according to 10 activity categories addressed by the classification algorithm. These categories were: lying, sitting, standing, active standing (sweeping the floor), walking, running, cycling, using a chest press fitness machine, using a leg press fitness machine, and rowing.

The acceleration pattern of each activity type was described by time- and frequency-domain features. These features provided information used by the classification algorithm to develop the knowledge necessary to identify activity types. For each sampling frequency, a number of features was calculated at each segment of 6.4 s of the acceleration signal (within each isolated activity task), from each sensing axis (Bonomi et al., 2009b).

Decision trees were developed by using the 5 sets of features measured in segments of the acceleration signal sampled at 20, 10, 5, 2.5, and 1.25 Hz. The purpose was to determine whether sampling frequencies lower than 20 Hz allowed the achievement of accurate classification performances, which was evaluated using the leave-one-subject-out cross-validation approach (C_V).

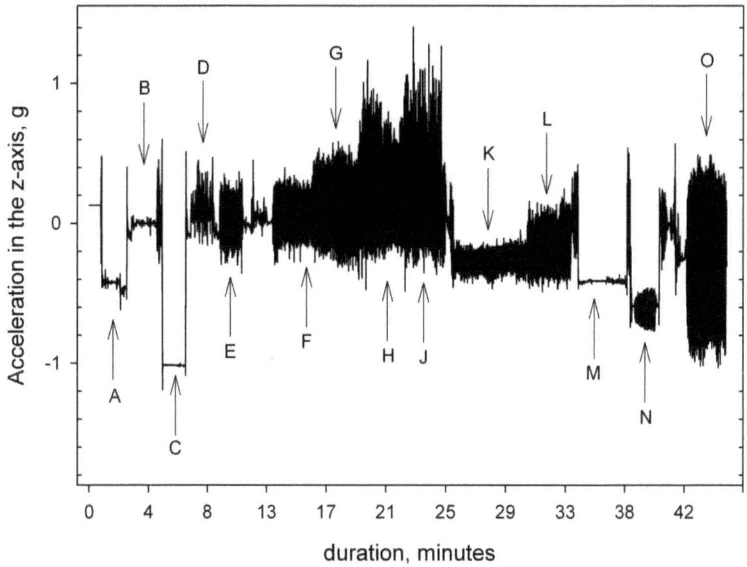

Fig. 3.2 Acceleration signal measured during the experimental protocol in the antero-posterior direction of the body (*Z axis*). *Arrows* highlight the signal measured during the tasks included in the test: (**a**) sitting; (**b**) standing; (**c**) lying; (**d**) sweeping the floor; (**e**) walking at self selected speed; (**f**) walking slowly on a treadmill; (**g**) walking fast on a treadmill; (**h**) running slowly on a treadmill; (**j**) running fast on a treadmill; (**k**) cycling at a slow pedaling rate; (**l**) cycling at a fast pedaling rate; (**m**) chest press; (**n**) leg press; (**o**) rowing

The findings were that the classification accuracy C_V of the decision tree decreased as the sampling frequency decreased (Fig. 3.3). However, no significant difference was observed between the C_V of the decision tree developed with the acceleration sampled at 20 Hz ($C_V = 81 \pm 12\%$) and that of the decision tree developed by sampling the acceleration at 10 and 5 Hz ($C_V = 79 \pm 11\%$, $p = 0.72$, and $C_V = 79 \pm 12\%$, $p = 0.65$, respectively). The C_V obtained at 20 Hz, at 10 Hz, and at 5 Hz was significantly higher than the accuracy measured at 2.5 Hz ($C_V = 69 \pm 14\%$, $p < 0.05$), and at 1.25 Hz ($C_V = 68 \pm 9\%$, $p < 0.05$). The C_V obtained at 10 Hz was not significantly different to that obtained at 5 Hz ($p = 0.95$). The C_V measured at 2.5 Hz and at 1.25 Hz were not significantly different ($p = 0.86$). Thus, use of a sampling frequency between 20 and 5 Hz led to similar classification performances.

Reducing the sampling frequency for the acquisition of the acceleration signal below 5 Hz resulted in a general decrease in the performance of the decision tree in identifying some activity types, such as walking, running, cycling, and rowing. This means that the information collected by the features was less able to describe the differences in the acceleration pattern of these activities, as the sampling frequency decreased. The reason can be determined by considering the Nyquist frequency – the frequency at which the signal presents the maximum frequency component – of the signal. According to the sampling theorem (or Shannon theorem) (Shannon,

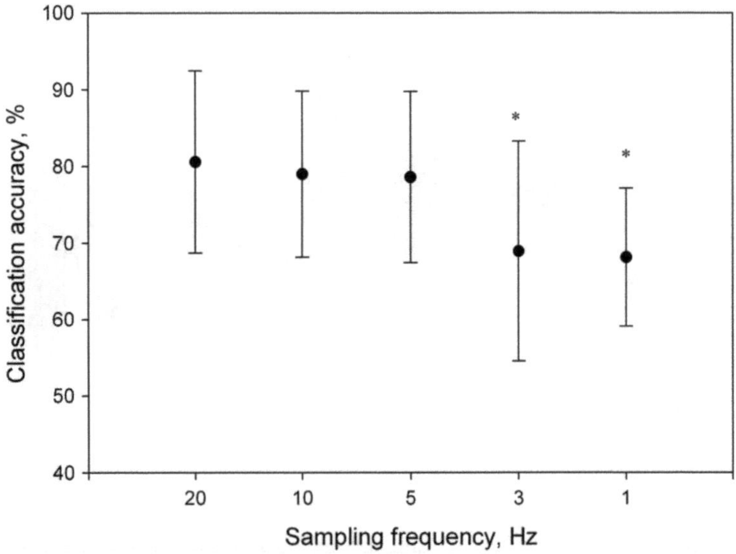

Fig. 3.3 Classification accuracy (C_V) of the decision tree developed for each sampling frequency according to leave-1-subject-out cross-validation; *, $p < 0.05$

1949), an acquisition system would not lose the information contained in the measured signal if the sampling frequency is equal to or higher than twice the Nyquist frequency. Since the aforementioned activities are highly cyclic and characterized by pronounced harmonic components, decreasing the sampling frequency below 5 Hz could have caused the loss of important data, as the Nyquist frequency of the acceleration signal could lie above half of the sampling frequency (Fig. 3.4). This might explain the reduced accuracy of the decision trees in classifying certain activity types when the sampling frequency was below 5 Hz. However, activities such as walking, running, and rowing when performed at a faster speed might produce harmonic components at higher frequencies than those observed in this study. In this case, a sampling frequency of 5 Hz might not be sufficient to describe the acceleration signal, as the Nyquist frequency of the measured signal might exceed 2.5 Hz (half of the sampling frequency).

In summary, acquiring the acceleration signal at the waist level using a sampling frequency of 5 Hz determines the achievement of classification performances comparable to that obtained using sampling frequencies of 10 and 20 Hz. Therefore, this specification represents a potential way to improve battery life and reduce power consumption for a wearable activity monitor used for recognizing generic categories of daily activities.

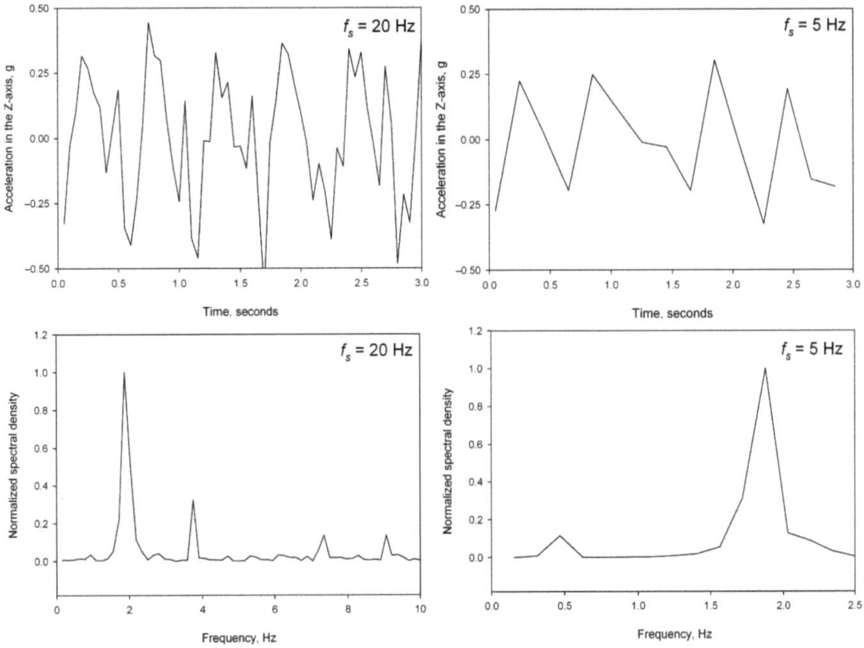

Fig. 3.4 Acceleration signal (*above*) and power spectral density (*below*) of the acceleration signal recorded using a sampling frequency of 20 Hz (*left*) or a sampling frequency of 5 Hz (*right*)

3.4 Segmentation of the Acceleration Signal

Classification algorithms identify activity types by evaluating attributes of the acceleration signal measured in portions of a defined length (segments). A segment of the acceleration includes a certain number of data points determined by the sampling frequency of the signal and by the time length of the segment. Given a certain sampling frequency, the longer the segment size the more samples are available for calculating attributes (features) of the acceleration. These acceleration features are used by classification algorithms to classify the type of an activity performed in a certain time interval. The use of short segments for the calculation of the acceleration features would improve the ability to correctly recognize short activities and to measure activity duration, supposing that the classification performance is constant regardless of the segment size. In literature the segmentation of the acceleration signal has been done using segments of 1 s (Zhang et al., 2003; Ermes et al., 2008b), 5.2 s (Ermes et al., 2008a), 6.7 s (Bao and Intille, 2004), or 15 s (Pober et al., 2006). However, a relationship has been observed between the classification accuracy of the decision tree classifiers and the length of the segment used to analyze the acceleration signal. A study of Bonomi et al. (2009b) investigated which segment size allowed the highest classification accuracy for identifying 7 activity types. Six

Table 3.1 Classification accuracy of the decision trees developed using different segment sizes

		F-score						
Segment size	C_V, %	Lie	Sit	Stand	AS	Walk	Run	Cycle
0.4 *s*	90.4*	100.0	85.7	53.9	67.6	97.3	99.1	89.3
0.8 *s*	91.9*	100.0	86.4	59.6	71.9	98.3	99.7	92.2
1.6 *s*	92.3*	100.0	86.6	58.0	72.8	98.8	99.9	93.3
3.2 *s*	92.6*	100.0	86.7	59.7	72.5	99.1	100.0	93.4
6.4 *s*	93.1	100.0	87.4	62.4	75.2	99.2	100.0	93.9
12.8 *s*	93.0	100.0	86.4	60.0	74.5	99.5	100.0	95.1

Segment size, length of the intervals used to segment the acceleration; C_V, average percentage of the correctly classified segments in the leave-one-subject-out cross-validation; F-score, is the harmonic mean of sensitivity and specificity of the classification method, and describes the ability of the decision tree in identifying each activity type; AS, active standing activity; *, significant difference ($p < 0.05$) as compared to C_V measured using segments of 6.4 s or 12.8 s.

decision trees were developed by analyzing the acceleration signal using segments of 0.4, 0.8, 1.6, 3.2, 6.4 and 12.8 s, including 8, 16, 32, 64, 128 and 256 samples at a sampling frequency of 20 Hz, respectively. The acceleration signal stored in each segment was processed to extract features in the time and frequency domain, and for each considered segment size a decision tree was developed. The findings were that decision trees developed using segments of 12.8 s ($C_V = 93.0\%$) and 6.4 s ($C_V = 93.1\%$) showed the highest classification accuracy, as tested using leave-one-subject-out cross-validation. The smaller the segment size considered, the lower the classification accuracy (Table 3.1). There was no significant difference between the classification accuracy of the models developed using segments of 12.8 and 6.4 s ($p = 0.41$). The paired t-test showed that the classification accuracies of the models developed using segments of 3.2 and 1.6 s were almost significantly different ($p = 0.05$).

Analyzing and classifying the acceleration signal with a high time resolution reduces the error for the definition of activity duration and increases the accuracy for the classification of short activities. The reason is that, using high resolution of analysis, the error introduced in the outcome of the classification algorithm due to activity transitions is minimized. When the segmentation of the acceleration signal is made by considering contiguous intervals, the use of short intervals increases the time resolution of analysis. However segments of a longer length might carry more meaningful information on the type of activity, improving the classification accuracy. Alternatively, a method employed to increase the time resolution is the segmentation of the signal in overlapping intervals (Bao and Intille, 2004). Using this technique, the time resolution is determined by the level of overlap between segments, and it can be increased without reducing the segment size. However, because of the overlap, misclassifications due to activity transitions would affect more segments, and thus, the propagation of the classification error occurs. Bonomi et al. (2009b) reported that the use of too short intervals for the computation of acceleration features led to a reduction of the classification accuracy. Using features measured in segments of 0.4 s reduced the classification accuracy by 3% compared to the one obtained with segments of 6.4 and 12.8 s. The decline of classification performance concerned most of the activity types. This can be explained by the fact that the features, when computed in shorter segments, were unable to fully represent the characteristics of a certain activity, and thus had a higher intra-class variability (variability within the same activity class), making the accuracy of decision trees lower.

For this reason, the choice of the segmentation technique should be carefully considered in order to develop algorithms with robust classification performances. Although the type of employed acceleration features and the type of chosen classification algorithms might influence the choice of the segment length, there is evidence that for daily physical activity, the use of segments shorter than 6 s introduces high intra-class variability in the acceleration features, which could reduce the classification accuracy.

3.5 Classification Algorithms and Data Recording

Several classification algorithms have been proposed to evaluate the acceleration features and to identify physical activity types. The most common classification algorithms are: threshold-based models, hierarchical models, decision trees, k-nearest neighbour, artificial neural network, support vector machine, Bayesian classifier, Markov models, and a combination of those (Preece et al., 2009). These models have been successfully employed to solve activity recognition problems, and the accuracy of these algorithms depended on several factors, such as the type of activities to be identified, the placement and number of sensors, the characteristics of the acquisition system, and the features considered. Therefore, the choice of the classification algorithm depends largely on the application of activity recognition.

When activity recognition is used to determine the daily engagement in physical activity, decision trees (Bao and Intille, 2004; Bonomi et al., 2009b; Ermes et al., 2008b), artificial neural networks (Zhang et al., 2003), and hidden Markov models (Pober et al., 2006) have been proposed to identify activity types. For this application the activity monitor is often designed to monitor physical activity for a long period of time, from a few weeks up to years. This means that the device should support activity recognition for a prolonged period of time. Traditionally, most of the processing steps necessary to recognize physical activity (calculation of the acceleration features, and identification of activity types) were performed off-line, for improving the battery life of the device. Indeed, limiting the computational time of the device's processing unit represents a major strategy to reduce power consumption. In this way, the activity monitor could simply store acceleration samples in the internal memory, and then determine off-line the features and the corresponding activity category (e.g. by using dedicated software on a personal computer). In this case the downloading of a few days of raw data from the activity monitor to the computer might result in a time consuming procedure. A possible solution could be to design activity monitors able to process the recorded samples on the device for feature calculation or, further, for identifying activity types. Thus, the activity monitor would only store, and eventually transfer, data in a more compact way, such as acceleration features or the types of activity performed. This strategy would imply the request of the use of simple acceleration features, and of simple classification algorithms which could be computed by the limited processing capacity of the internal CPU of an activity monitor.

In consequence, reducing the amount of data stored by the activity monitor implies the use of features of easy computation, e.g. time domain features, and of classification algorithms of easy implementation, such as threshold-based models, hierarchical models, or decision trees. In this way, completing most of the processing steps necessary for activity recognition in the processing unit of the device will improve the user friendliness of the activity monitor. Furthermore, this on-board approach of activity recognition improves the ability of the activity monitor to interact with the behavior of the user, which is the ultimate goal in context-aware applications. However, one should carefully consider that classification accuracy could decrease by using too simple classification algorithms to process the acceleration signal.

Acknowledgement The author thanks Chen Xin for recruiting the study participants and for following the experimental protocol of the study aimed at determining the optimal sampling frequency of accelerometers for activity recognition.

References

Antonsson, E.K., Mann, R.W.: The frequency content of gait. J. Biomech. **18**, 39–47 (1985)

Bao, L., Intille, S.S.: Activity recognition from user-annotated acceleration data. Pervasive Comput. Proc **3001**, 1–17 (2004)

Boissy, P., Choquette, S., Hamel, M., Noury, N.: User-based motion sensing and fuzzy logic for automated fall detection in older adults. Telemed. J. E. Health **13**, 683–693 (2007)

Bonomi, A.G., Plasqui, G., Goris, A.H., Westerterp, K.R.: Improving assessment of daily energy expenditure by identifying types of physical activity with a single accelerometer. J. Appl. Physiol. **107**(3), 655–661 (Sept. 2009a). Epub 25 June 2009

Bonomi, A.G., Goris, A.H., Yin, B., Westerterp, K.R.: Detection of type, duration, and intensity of physical activity using an accelerometer. Med. Sci. Sports Exerc. **41**(9), 1770–1777 (Sept. 2009b)

Brage, S., Wedderkopp, N., Ekelund, U., Franks, P.W., Wareham, N.J., Andersen, L.B., Froberg, K.: Features of the metabolic syndrome are associated with objectively measured physical activity and fitness in Danish children: the European Youth Heart Study (EYHS). Diabetes Care **27**, 2141–2148 (2004)

Crouter, S.E., Churilla, J.R., Bassett, D.R., Jr.: Estimating energy expenditure using accelerometers. Eur. J. Appl. Physiol. **98**, 601–612 (2006)

Ekelund, U., Brage, S., Besson, H., Sharp, S., Wareham, N.J.: Time spent being sedentary and weight gain in healthy adults: reverse or bidirectional causality? Am. J. Clin. Nutr. **88**, 612–617 (2008)

Ermes, M., Parkka, J., Cluitmans, L.: Advancing from offline to online activity recognition with wearable sensors. Engineering in medicine and biology society: 30th annual international conference of the IEEE, pp. 4451–4454 (2008a)

Ermes, M., Parkka, J., Mantyjarvi, J., Korhonen, I.: Detection of daily activities and sports with wearable sensors in controlled and uncontrolled conditions. IEEE Trans. Inform. Technol. Biomed. **12**, 20–26 (2008b)

Gurley, R.J., Lum, N., Sande, M., Lo, B., Katz, M.H.: Persons found in their homes helpless or dead. N. E. J. Med. **334**, 1710–1716 (1996)

Harris, T.J., Owen, C.G., Victor, C.R., Adams, R., Ekelund, U., Cook, D.G.: A comparison of questionnaire, accelerometer, and pedometer: measures in older people. Med. Sci. Sports Exerc. **41**, 1392–1402 (2009)

Haskell, W.L., Lee, I.M., Pate, R.R., Powell, K.E., Blair, S.N., Franklin, B.A., et al.: Physical activity and public health: Updated recommendation for adults from the American College of Sports Medicine and the American Heart Association. Med. Sci. Sports Exerc. **39**, 1423–1434 (2007)

Hu, F.B., Li, T.Y., Colditz, G.A., Willett, W.C., Manson, J.E.: Television watching and other sedentary behaviors in relation to risk of obesity and type 2 diabetes mellitus in women. JAMA **289**, 1785–1791 (2003)

Karantonis, D.M., Narayanan, M.R., Mathie, M., Lovell, N.H., Celler, B.G.: Implementation of a real-time human movement classifier using a triaxial accelerometer for ambulatory monitoring. IEEE Trans. Inf. Technol. Biomed. **10**, 156–167 (2006)

Mathie, M.J., Coster, A.C., Lovell, N.H., Celler, B.G.: Detection of daily physical activities using a triaxial accelerometer. Med. Biolog. Eng. Comput. **41**, 296–301 (2003)

Melanson, E.L., Jr., Freedson, P.S.: Physical activity assessment: a review of methods. Crit. Rev. Food Sci. Nutr. **36**, 385–396 (1996)

Plasqui, G., Westerterp, K.R.: Physical activity assessment with accelerometers: an evaluation against doubly labeled water. Obesity **15**, 2371–2379 (2007)

Pober, D.M., Staudenmayer, J., Raphael, C., Freedson, P.S.: Development of novel techniques to classify physical activity mode using accelerometers. Med. Sci. Sports Exerc. **38**, 1626–1634 (2006)

Preece, S.J., Goulermas, J.Y., Kenney, L.P., Howard, D., Meijer, K., Crompton, R.: Activity identification using body-mounted sensors–a review of classification techniques. Physiol. Meas. **30**(4), R1–R33 (2009 Apr.). Epub 2 Apr. 2009

Shannon, C.E.: Communication in the presence of noise. Proc. Inst. Radio Eng. **37**(1), 10–21 (1949)

Veltink, P.H., Bussmann, H.B., de Vries, W., Martens, W.L., Van Lummel, R.C.: Detection of static and dynamic activities using uniaxial accelerometers. IEEE Trans. Rehabil. Eng. **4**, 375–385 (1996)

Wild, D., Nayak, U.S., Isaacs, B.: How dangerous are falls in old people at home? Br. Med. J. (Clin. Res. Ed) **282**, 266–268 (1981)

Zhang, K., Werner, P., Sun, M., Pi-Sunyer, F.X., Boozer, C.N.: Measurement of human daily physical activity. Obes. Res. **11**, 33–40 (2003)

Ermes, M., Pärkkä, J., Mantyjarvi, J., Korhonen, I.: Detection of daily activities and sports with wearable sensors in controlled and uncontrolled conditions. IEEE Trans. Inf. Technol. Biomed. 12, [illegible] (2008)

[illegible]

Harris, T.J., Owen, C.G., Victor, C.R., Adams, R., Ekelund, U., Cook, D.G.: A comparison of questionnaire, accelerometer, and pedometer: measures in older people. Med. Sci. Sports Exerc. 41, 1392–1402 (2009)

Haskell, W.L., Lee, I.M., Pate, R.R., Powell, K.E., Blair, S.N., Franklin, B.A., et al.: Physical activity and public health: updated recommendation for adults from the American College of Sports Medicine and the American Heart Association. Med. Sci. Sports Exerc. 39, 1423–1434 (2007)

Hu, F.B., Li, T.Y., Colditz, G.A., Willett, W.C., Manson, J.E.: Television watching and other sedentary behaviors in relation to risk of obesity and type 2 diabetes mellitus in women. JAMA 289, 1785–1791 (2003)

Karantonis, D.M., Narayanan, M.R., Mathie, M., Lovell, N.H., Celler, B.G.: Implementation of a real-time human movement classifier using a triaxial accelerometer for ambulatory monitoring. IEEE Trans. Inf. Technol. Biomed. 10, 156–167 (2006)

Mathie, M.J., Coster, A.C., Lovell, N.H., Celler, B.G.: Detection of daily physical activities using a triaxial accelerometer. Med. Biol. Eng. Comput. 41, 296–301 (2003)

Melanson, E.L., Jr., Freedson, P.S.: Physical activity assessment: a review of methods. Crit. Rev. Food Sci. Nutr. 36, 385–396 (1996)

Plasqui, G., Westerterp, K.R.: Physical activity assessment with accelerometers: an evaluation against doubly labeled water. Obesity 15, 2371–2379 (2007)

Pober, D.M., Staudenmayer, J., Raphael, C., Freedson, P.S.: Development of novel techniques to classify physical activity mode using accelerometers. Med. Sci. Sports Exerc. 38, 1626–1634 (2006)

Preece, S.J., Goulermas, J.Y., Kenney, L.P., Howard, D., Meijer, K., Crompton, R.: Activity identification using body-mounted sensors—a review of classification techniques. Physiol. Meas. 30(4), R1–R33 (2009 Apr), Epub 2 Apr 2009

Shannon, C.E.: Communication in the presence of noise. Proc. IRE 37(1), 10–21 (1949)

[illegible]

[illegible]

[illegible] physical activity. Obes. Res. 11, 33–40 (2003)

Chapter 4
The Use of Psychophysiological Measures During Complex Flight Manoeuvres – An Expert Pilot Study

Wolfram Boucsein, Ioana Koglbauer, Reinhard Braunstingl, and K. Wolfgang Kallus

Abstract Simulator training is common for commercial pilots but not in general aviation. Since unusual flight attitudes, stalls and spins account for about one third of fatal accidents in pilots flying according to visual flight rules, the present authors are currently evaluating a simulator training program developed for this group of pilots. Our study does not only use the progress in recovering from unusual manoeuvres as criterion for training success, but also psychophysiological recordings during the actual flight manoeuvres. Based on a theoretical arousal/emotion brain model (Boucsein and Backs, 2009), heart rate, heart rate variability and various electrodermal parameters were chosen for in-flight recording in an aerobatic plane (Pitts S-2B), flown by an expert aerobatic pilot who will be the flight instructor during the test flights before and after simulator training. In the present study, psychophysiological recordings were taken before, during and after flying into and recovering from extreme pitch, overbanking, power-off full stall and spin. To control for the influence of high acceleration, G-forces were recorded by an inertial platform. Results from our expert pilot study demonstrate the usability of psychophysiological measures for not only determining stress/strain processes, but also different kinds of arousal, i.e., general arousal, preparatory activation and emotional arousal, during complex flight manoeuvres.

4.1 Introduction

Pilot failures in recovery to straight-and-level flight from unusual attitudes and stall/spin manoeuvres essentially contributed to fatal aviation accidents in the last decades. In the commercial aviation 2,100 people lost their lives in 32 accidents resulted of unusual attitudes between 1994 and 2003 (Boeing Commercial Airplane Group, 2004). Although no appropriate statistics exist for general aviation, the AOPA Air Safety Foundation (2003) revealed for the period between 1991 and 2000

W. Boucsein (✉)
Physiological Psychology, University of Wuppertal, Wuppertal, Germany
e-mail: boucsein@uni-wuppertal.de

J. Westerink et al. (eds.), *Sensing Emotions*, Philips Research Book Series 12,
DOI 10.1007/978-90-481-3258-4_4,

that stall/spin accidents accounted for 10% of the general aviation accidents and highest fatality rating (30%). Interestingly, even flight instructors are not prevented from being caught in this type of accidents, since 91% of the investigated 44 instructional stall/spin accidents occurred during dual instruction, and only 9% during solo training. While causes of incidents and/or accidents cannot be entirely controlled, since they are found in the whole environment-aircraft-pilot system, it is essential that even experienced general aviation pilots ought to be trained in re-establishing safe flight. Therefore, the general aim of our current research is to develop and probe a training procedure for this group of pilots, including extended ground briefing and simulator training of unusual attitude and stall/spin awareness and recoveries.

Besides technical performance features to be included in the process of developing the training and evaluation procedure, the present study attempted to give insight into non-technical strategies and adaptive psychophysiological arousal and emotion regulation during different cognitive, emotional and physical demands of the above mentioned flight tasks. As a theoretical framework of analyzing non-technical strategies, we used the model of the "anticipation-action-comparison unit" (Kallus et al., 1997) and the concept of "situation awareness loop" (cf. Kallus and Tropper, 2007).

The model of the "anticipation-action-comparison unit" considers predictions of the future mental picture based on key elements of the current situation, mental models and previous experience. Actions are anticipated envisioning their effects, and feedback comparisons of predicted and actual effects close the loop of task management. Anticipatory processes do not only involve different levels of information processing, but take also place on different levels of central nervous organization, starting from unconscious anticipatory eye-movements and ending with complex conscious planning processes (Kallus and Tropper, 2006). Since processes of anticipation, preparation, planning and action may involve different neurophysiological systems, they can be objectively identified by appropriate psychophysiological indicators (Boucsein and Backs, 2009).

The concept of "situation awareness loop" integrates theories of situation awareness (Endsley, 1988) and anticipatory behavioral control (Hoffmann, 1993; 2003). Situation awareness is defined as the perception of environmental cues, the comprehension of their meaning, and the projection of their status in the near future. Kallus (2009) proposed that anticipatory processes are of much higher importance for flying performance than classical approaches of situation awareness presume, since they are not only a result of perceptual processes but also have the property to influence perception itself. Therefore, Kallus fostered including the sequence of anticipation, perception, comprehension and projection in the near future into a feedback loop, thus enabling anticipatory behavioral control. This concept, which was originally formed in the domain of cognitive psychology, has been more recently brought into connection with neurophysiological methodology (Pezzulo et al., 2007).

The use of psychophysiological concepts for determining anticipatory processes is not new to the field of flight psychology. Kallus and Tropper (2007) assessed the role of anticipative processes for military jet pilots executing critical flight manoeuvres in a motion based flight simulator. They analyzed heart rate (HR) and heart rate variability (HRV) in different sections of a "black hole approach" manoeuvre by pilots distributed ex post in three performance groups: crash, problems and landing. All pilots showed an increased HR at the beginning of their flight profile, the increase being much higher by pilots who crashed later. These findings indicate different anticipatory processes of the three performance groups, the anticipatory increase of psychophysiological arousal by pilots who crashed being higher as compared to that of pilots who landed. Pilots with good landings showed higher HRV at the beginning of the manoeuvre and smaller HRV values about 40 s before touchdown as compared to pilots who crashed. These results could be replicated in a second study with private and professional pilots, where mean HR of pilots who crashed later were increased from the beginning of the manoeuvre as compared to pilots who landed safely. Pilots of the crash group seemed to anticipate problems more or less consciously but continued the erroneous approach instead of performing a go-around.

Boucsein (2007) gave an example of how psychophysiological responses during the performance of complex flight manoeuvres could be interpreted in a neurophysiological framework, which was first developed by Boucsein (1992). In a case study with himself as pilot he recorded cardiac and electrodermal activity during twenty eight segments of a flight. Psychophysiological parameters could not only differentiate between flight segments such as take off, climb, manoeuvring flight, approach and landing. In addition, anticipation and performance of complex flight manoeuvres such as stalls and steep turns revealed characteristic psychophysiological patterns, which were interpreted within the above mentioned neurophysiological arousal/emotion model as affect arousal, effort, preparatory activation and general arousal (Boucsein and Backs, 2009). This model synthesizes empirically proved connections of arousal and information processing mechanisms associated with emotional and motivational influences, and how they affect central and peripheral psychophysiological parameters. Cardiac and electrodermal measures proved to be sensitive indicators of mental, physical and emotional load (Boucsein and Backs, 2000). The present study used this model to determine the contribution of different kinds of arousal during anticipation, onset, recovery and repositioning phases of unusual flight attitudes (extreme pitch and overbanking), stalls and spins.

Since all these flight manoeuvres involve G-forces considerably exceeding the natural ones, it was important to control the influences of the additional G-force on the pilot's cardiac activity. It was demonstrated, in the realm of aerospace medicine studies with the centrifuge, that human cardiovascular activity is significantly influenced by positive acceleration on the head-to-feet axis (+z-axis in body fixed coordinates). Burton and Whinnery (1996) reported cardiovascular G-effects from six subjects that were exposed to a 1 Gz control condition and several G-conditions above the natural level (+2, +3, and +4 Gz). Exposure to more than +2Gz without using an anti-G suit resulted in a decrease of cardiac output, an increase of HR, a decrease of the stroke index, an increase of heart arterial pressure, an increase of vascular resistance and a reduction of arterial oxygen saturation. Burton and Whinnery (1996, pp. 208–209) concluded, that "the accelerative force effect on heart rate is primarily a response to the baroreceptor cardiovascular compensatory reflex to a reduced arterial blood pressure (Pa) at the site of the carotid sinus and the decrease in cardiac output". To enable control of possible G-effects on cardiac activity, G-forces were recorded in parallel to physiological measures in the present study.

As a preparation for designing our training with general aviation pilots, we considered the expert pilot, who was the flight instructor during the test flights before and after simulator training, being a suitable model for successful technical and psychophysiological performance in the to-be-trained flight manoeuvres (Koglbauer et al., 2011).

4.2 Methods

The third author – an expert aerobatic pilot and licensed aerobatic flight instructor – performed the following manoeuvre sequence twice consecutively in a single solo flight: extreme pitch, overbanking, power-off full stall and spin with two rotations. During the entire flight, electrocardiogram (ECG) and electrodermal activity (EDA) were recorded, together with flight mechanical data of the aircraft such as Euler angles and acceleration forces of three body fixed axes.

The expert pilot identified four phases of flight manoeuvre management: anticipation, manoeuvre onset, recovery and repositioning. During the anticipation phase, the pilot reported to have mentally activated the manoeuvre scenario "what I have to do and how to do it" in a precise manner. During onset, he initiated characteristic angular and speed parameters of the manoeuvre, and during recovery, he established initial horizontal flight attitude and airspeed. In the repositioning phase, he re-established location and altitude for the next manoeuvre.

4.2.1 In-Flight Recordings

Recordings of chest ECG, respiration and EDA as skin conductance from the left hand, thenar/hypothenar, were performed with the Varioport system (distributed by Becker Meditec, Karlsruhe, 2005). Baselines of 2 min were obtained before and after the flight. In-flight mechanical data were recorded in the Body Fixed Coordinate System, by means of an inertial platform developed by Graz University of Technology, including an aviation-certified laser gyro, a MEMS gyro, a GPS sensor, and acceleration sensors for each of the three axes. The Body Fixed Coordinate System has its origin in the aircraft's mass centre and the axes x, y and z are fixed in and move with the aircraft. Recordings in this system provide objective flight data for the dynamic phenomena that pilots perceive during flight, such as Euler angles and accelerations of the three axes. To obtain exact points of time for the beginning and end of each phase, the Body Fixed Coordinate System data from the entire flight were reproduced in a simulation interface which provided both instrument displays and outside views of the aircraft. Sequences of the flight relevant for this study were cut out of the simulation and recorded in a specifically developed data-logger, providing numerical mechanical data of Euler angles and accelerations with a resolution of 10 Hz. Temporal sequences of all phases were matched with ECG and EDA recordings for processing, so that objective information regarding flight performance, G-load, plus the pilot's cardiac and electrodermal arousal was available for each sequence of the flight. Psychological and somatic aspects of psychophysiological strain were assessed by questionnaires before and after the flight, and workload was evaluated after the flight by the NASA-TLX.

4.2.2 Data Evaluation

Independent variables were considered the type of flight manoeuvre to perform, i.e., extreme pitch, overbanking, stall and spin, as well as the four different task phases that constituted each manoeuvre: anticipation, onset, recovery and repositioning. Phases of the manoeuvres in-flight were also compared to resting periods before and after the flight session.

From the ECG, the following psychophysiological parameters were taken: mean HR, HRV as standard deviation (SD) and as mean square of successive differences (MSSD), raw inter-beat intervals (IBIs) and IBI variability as SD and as MSSD. For EDA, skin conductance level (SCL), non-specific skin conductance response frequency (NS.SCR freq.), mean amplitude of SCRs (SCR amp.) and mean recovery time of SCRs (SCR rec.t.) were evaluated using the program EDA-Vario (Schaefer, 2007, Version 1.8). Subjective physical strain was assessed by a Physical Symptom Check List (Erdmann and Janke, 1978, MKSL-24-ak, Mehrdimensionale Körperliche Symptomliste, unpublished), subjective psychological strain was evaluated by the Brief Adjective Check List (Janke et al., 1986, BSKE (EWL)-ak, Befindlichkeitsskalierung nach Kategorien und Eigenschaftswörtern, unpublished), and NASA-TLX (Hart and Staveland, 1988) was used as measure of subjective workload.

Statistical evaluation of the differences between the anticipation, onset, recovery and repositioning phases was performed by means of Friedman test (asymmetric significance). Individual comparisons between the anticipation phase and the other phases including resting values were made by Wilcoxon tests, and Spearman correlations were calculated. To evaluate the possible influence of G-forces on the ECG, raw IBIs and acceleration of the three axes expressed in g (1 g = 9.81 m/s^2) were analyzed for the phases of flight with G-forces above the natural ones, i.e., during maneuvre onset and recovery.

4.3 Results

After the flight, the expert pilot evaluated his effort as moderate and the flight as "fun". He attempted to perform the manoeuvres precise, clear and relatively slow. He reported an anticipation phase of 5–7 s, during which he mentally activated the manoeuvre scenario, which he described as "thinking what I have to do and how to do it". In addition, he described having processed several anticipation-action-comparison units in parallel during the other three phases of each flight task. Comparisons of subjective physical strain ratings of the expert pilot before and after the flight, assessed by the Physical Symptom Check List, indicated a slight decrease of relaxation and an increase of physical strain. Subjective psychological ratings, evaluated by the Brief Adjective Check List, showed stable high levels of good mood and vigilance, a slight decrease of introversion and a slight increase of anxiety from the beginning to the end of the flight. Post-flight NASA-TLX ratings indicated low mental, physical and temporal demands as well as low effort and frustration.

4.3.1 Analysis of Psychophysiological Parameters

Since each of the four flight manoeuvres was flown twice, the statistical evaluation is based on eight data sets. For each psychophysiological parameter, differences between all four phases of the flight tasks were tested with Friedman and Wilcoxon non-parametric tests ($N = 8$, $df = 3$). Non-parametric correlations were calculated with Spearman's ρ.

Significant differences were obtained for HR ($\chi^2 = 8.100$, $p = 0.044$), for HRV calculated MSSD ($\chi^2 = 13.050$, $p = 0.005$) and as SD ($\chi^2 = 13.050$, $p = 0.005$), for SCL ($\chi^2 = 12.266$, $p = 0.007$) and mean SCR amp. ($\chi^2 = 9.117$, $p = 0.028$), but not for SCR rec.t. ($\chi^2 = 7.622$, $p = 0.055$) and NS.SCR freq. ($\chi^2 = 7.350$, $p = 0.062$). Means and standard deviations for all ECG and EDA parameters, averaged over the two repeatedly flown manoeuvres, are given in Table 4.1. Mean HR increased significantly from the resting phases to the anticipation phases ($z = -2.089$, $p = 0.037$) and further during recovery ($z = -2.380$, $p = 0.017$), being lower, (but not significantly) during repositioning and manoeuvre onset. The SCL significantly increased in the anticipation phases compared to the resting phases ($z = -1.828$, $p = 0.068$)

Table 4.1 Means and standard deviations (SD) for all parameters extracted. "SCR" refers to NS.SCRs. SCR freq. refers to 1 min; units for SCR amp. are μS, while SCR rec.t. is given in s; both HRVs are in arbitrary units

Phases parameters	Resting		Anticipation		Onset		Recovery		Repositioning	
	Mean	SD	Mean	SD	Mean	SD	Mean	SD	Mean	SD
HR [bpm]	84.39	14.29	108.42	4.66	108.19	4.11	114.60	5.96	111.91	7.35
HRV(MSSD)	4.70	1.07	1.47	0.74	2.04	0.97	3.61	0.83	4.53	2.92
HRV(SD)	16.32	12.68	0.85	0.38	2.20	0.95	2.75	1.45	2.88	1.64
SCL [μS]	12.53	8.24	18.89	0.55	19.01	0.62	19.38	0.78	18.89	0.57
SCR freq.	57.00	14.84	37.50	15.83	35.01	5.37	25.08	6.72	35.59	12.50
SCR amp.	0.03	0.01	0.06	0.03	0.08	0.04	0.20	0.08	0.08	0.05
SCR rec.t.	0.21	0.09	0.22	0.09	0.12	0.05	0.25	0.07	0.22	0.07

and further during recovery ($z = -2.380$, $p = 0.017$), but not in the onset and repositioning phases.

The HRV (MSSD) decreased significantly from the resting phases to the anticipation phases ($z = -2.089$, $p = 0.037$) and was significantly lower compared to onset ($z = -2.521$, $p = 0.012$), recovery ($z = -2.380$, $p = 0.017$) and repositioning ($z = -2.521$, $p = 0.012$) phases. HRV (SD) was also significantly lower during anticipation compared to the resting phases ($z = -2.089$, $p = 0.037$) and again increased significantly in the recovery ($z = -2.521$, $p = 0.012$) and during repositioning ($z = -2.521$, $p = 0.012$), but not in the onset phases. The NS.SCR freq. continuously diminished from the resting phases through anticipation, onset and recovery, increasing again during repositioning, but the differences reached significance only between the anticipation and recovery phases ($z = -1.960$, $p = 0.05$). Mean NS.SCR amp. reached its peak during recovery, which was significant compared to the resting phases ($z = -2.375$, $p = 0.018$), but not to the other phases. Mean NS.SCR rec.t. was significantly lower in the onset phases compared to the resting phases ($z = -1.963$, $p = 0.05$), while the differences to the other phases did not reach significance. There was a positive correlation between HR during anticipation and onset ($\rho = 0.786$, $p = 0.021$). The SCL values of the anticipation phase correlated positively with those of the onset ($\rho = 0.929$, $p = 0.001$), recovery ($\rho = .833$, $p = 0.010$) and repositioning ($\rho = 0.905$, $p = 0.002$) phases. Correlations of the remaining parameters between anticipation and the other phases of task did not reach significance.

4.3.2 Analysis of Cardiac Activity in Phases of Flight with High Acceleration

To determine whether the cardiac activity was influenced by high-G acceleration, raw IBI data were plotted together with accelerations in all three axes. Since the resolution of acceleration data was 10 Hz, different scales of the abscissa had to be applied. Figure 4.1 shows that raw IBIs during the different phases of a spin with two rotations yielded a similar pattern for the first (solid line) and the second spin (dashed line). The decrease of IBIs during anticipation points to an influence of higher mental and physical effort, which was presumably not too much influenced by G-forces, since acceleration in the z-direction stayed as low as +1 Gz until 0.5 s before the end of the interval between onset and recovery (solid and dotted black lines in Figure 4.2).

Thereafter, G-forces increased to 1.5 Gz, which did not seem to considerably influence the IBIs (Fig. 4.1). Not earlier than in the middle of the recovery phase, Gz-forces accelerated to more than +2.5 Gz, which may have considerably contributed to the diminishing IBIs at the end of the recovery phase. Given the present restrictions in space, IBI and G-forces diagrams for the other three manoeuvres cannot be shown here. However, the diminishing influence of high Gz-forces on IBIs towards the end of the manoeuvre was also seen during recovery from the power-off

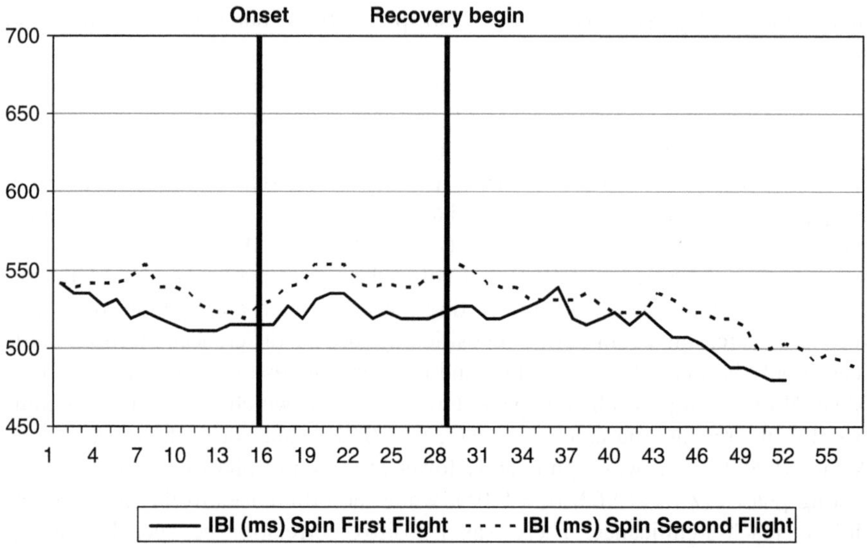

Fig. 4.1 Number and length of IBIs during anticipation, onset and recovery of the two-rotation spin manoeuvres

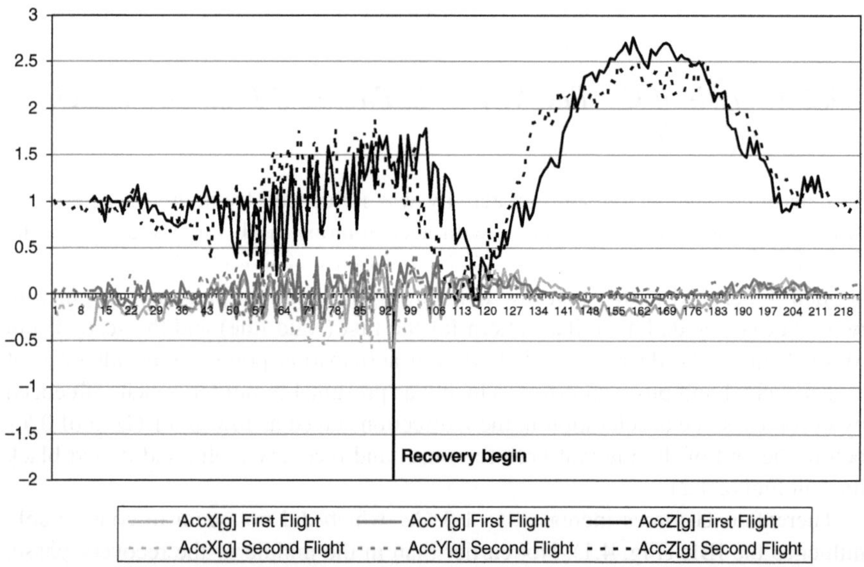

Fig. 4.2 Accelerations during the two-rotations spins (10 Hz resolution). Body fixed coordinates: +x = forward, +y = right, +z = down

full stall and the overbanked attitude where +2 Gz were reached, but not in recovering from the extreme pitch, where the Gz-force did not exceed +1.5 g. At first glance, it looks as if accelerations of 2 Gz and more, but not those below 2 Gz, considerably influence cardiac activity. However, during the two overbanking manoeuvre, forces up to +3 Gz acted upon the expert pilot, but IBIs did not fall below 550 ms.

Because of the relatively low number of observations, only rather preliminary statistical analyses of IBI parameters could be performed by means of the Wilcoxon Z test. The IBIs were lower in the recovery compared to the onset phases ($Z = -2.1$, $p = 0.036$). IBI variability calculated as SD was significantly lower in the onset phases ($Z = -2.1$, $p = 0.036$), a tendency that was also reflected in the IBI variability calculated as MSSD, but did not reach significance. Two-tailed correlation tests performed with Spearman's ρ were not significant for the mean IBI: $\rho = 0.143$ and IBI variability (SD): $\rho = -0.214$. However, IBI variability (MSSD) showed strong correlations between these two phases: $\rho = 0.738$, $p = 0.037$.

4.4 Discussion

According to the theoretical framework given in the introduction, our primary interest focused on the amounts of the expert pilot's psychophysiological arousal/emotion in the anticipation phases, in comparison with the resting periods before and after the flight and with the three other task phases. The differences will be interpreted according to the implications of Boucsein's (1992) four-arousal model, the extended version of which was published in Boucsein and Backs (2009), for changes in autonomic nervous system measures, given in their table 35.1. The marked and significant increase of HR and SCL during the anticipation phases compared to resting primarily reflects an increase of general arousal, which has to do with the task being performed during actual flying. Both measures further increase during recovery, possibly reflecting amplified actions necessary in these phases. The similarity of the indicator functions of both measures is also reflected in the rather high and significant correlations between anticipation and onset in HR and SCL, whereas correlations between anticipation and the other three flight phases reached a significant height.

The anticipation phases (5–7 s) in which the expert pilot mentally activated the manoeuvre scenario is not only characterized by an increase of HR but also by a significant decrease in both HRV measures, reflecting the amount of mental effort needed plus the activation of Broadbent's "higher level" cognitive system (see Boucsein and Backs, 2000), but possibly also a behavioral inhibition component, characterizing mental workload that comes without overt responses. During the following three phases of flight, HRV increases again significantly, indicating relaxation during recovery and repositioning. This increase is not yet significant during manoeuvre onset, which is complemented by the only significant decrease of SCR rec.t. during all phases of flight, indicating together that the manoeuvre onsets are still mentally demanding (Boucsein, 1992).

Interestingly, the expert pilot reported positive emotions and relative low subjective strain associated with performing the highly complex flight manoeuvres. The only sight increase of anxiety is reflected in the lack of significant increases in nonspecific electrodermal responses, which would have otherwise reflected a negatively tuned affective response. In contrast, NS.SCR freq. recordings from a non-expert pilot (Boucsein, 2007) flying stalls and steep turns clearly indicated an increased affect arousal during these manoeuvres. The expert pilot observed in the present study yielded a continuous decrease of NS.SCR freq. from resting over anticipation and onset to recovery and only a slight increase thereafter. The latter could also be a movement artifact since the hand wearing the EDA electrodes had to be used for the throttle. The NS.SCR freq. being significantly lower in the recovery phase compared to the anticipation phase supports the relaxing property of recoveries that was already seen in HRV. The observation that the mean NS.SCR amp. reaches its peak during the recovery phase points to an increase of cognitive activity together with preparatory activation in this phase.

The simultaneous analysis of flight mechanical and psychophysiological data recorded during the flight does not reveal systematic effects of Gz-strain on the pilot's cardiac activity. Even Gz-forces as high as between +2.5 and +3 g do not systematically diminish IBIs, although precaution should be applied when G-forces exceed +2 Gz. For the first time in aviation psychophysiology, influences of G-forces on cardiac activity were not only probed for IBIs, but also for IBI variability. No evidence for mixed acceleration within phase effects on the IBI variability (SD and MSSD) is found. However, significant correlations are obtained between IBI values (evaluated as MSSD) of phases with high acceleration. These results might be specific for the expert pilot, who generally associates high-G experience with positive emotions and automatically counteracts negative effects of acceleration by using anti-G straining manoeuvres.

In conclusion, our results give an insight into the adaptive psychophysiological arousal and emotion regulation of an expert pilot during different cognitive, emotional and physical demands of real flight tasks. It was demonstrated that recordings of cardiac and electrodermal activity can be performed during complex flight manoeuvres in an aerobatic plane without data loss. Even rather high G-forces did not obscure the diagnosticity and specificity of psychophysiological parameters for different kinds of arousals (Boucsein and Backs, 2009). Thus, psychophysiological recording as used in the present study turned out to be suitable as a model for the real test flights which was performed before and after simulator training for general aviation pilots flying according to visual flight rules by Koglbauer et al. (2011). Furthermore, our study of expert flight task management showed us that safe performance does not only manifest within boundaries of the flight task itself. We conclude that anticipation is not only a matter of situation awareness but additionally comprises a comparison of the actual flight situation with its expected changes. Anticipatory processes and post-task echoing of the flight task in the post-recovery phase are of great diagnostic value for safety relevant non-technical skills. Following the line of the present study, both anticipation as well as a post-recovery phase will be explicitly introduced in our future flight training and performance evaluation methodology. Furthermore, the content of the to-be-trained manoeuvre procedures will be split into distinct anticipation-action-comparison units (Kallus et al., 1997), to be visualized for the trainee as early as during ground briefing, fostering a sequence of perception, comprehension and projection in the pilot (Kallus, 2009).

Acknowledgement This research was supported and financed by the Austrian Research Promotion Agency (FFG) and the company AMT Aircraft Manufacturing Technology GmbH, Austria.

References

AOPA Air Safety Foundation: Stall/Spin: Entry point for crash and burn? In: Air Safety Foundation Special Reports, Maryland. Available online: http://www.aopa.org/asf/publications/topics/stall-spin.pdf (2003)

Boeing Commercial Airplane Group: Statistical Summary of Commercial Jet Airplane Accidents, Worldwide Operations, 1994–2003. Airplane Safety Engineering, Seattle, WA (2004)

Boucsein, W.: Electrodermal Activity. Plenum Press, New York, NY (1992)

Boucsein, W.: Psychophysiologie im Flug – eine Fallstudie. In: Richter, P.G., Rau, R., Mühlpfordt, S. (eds.) Arbeit und Gesundheit. Zum aktuellen Stand in einem Forschungs- und Praxisfeld, pp. 130–142. Pabst Science Publishers, Lengerich (2007)

Boucsein, W., Backs, R.W.: Engineering psychophysiology as a discipline: historical and theoretical aspects. In: Backs, R.W., Boucsein, W. (eds.) Engineering Psychophysiology. Issues and Applications, pp. 3–30. Lawrence Erlbaum Associates, Mahwah, NJ (2000)

Boucsein, W., Backs, R.W.: Methods and models for the psychophysiology of emotion, arousal, and personality. In: Duffy, V.G. (ed.) Handbook of Digital Human Modeling, pp. 35-1–35-18. CRC Press/Taylor & Francis, Boca Raton, FL (2009)

Burton, R.R., Whinnery, J.E.: Biodynamics: sustained acceleration. In: DeHart, R.L. (ed.) Fundamentals of Aerospace Medicine, 2nd edn, pp. 201–260. Williams & Wilkins, Baltimore, MD (1996)

Endsley, M.R.: Design and evaluation for situation awareness enhancement. Proceedings of the Human Factors Society 32nd Annual Meeting, vol. 1, pp. 97–101. Human Factors Society, Santa Monica, CA (1988)

Hart, S.G., Staveland, L.E.: Development of NASA-TLX (Task Load Index): results of empirical and theoretical research. In: Hancock, P.A., Meshkati, N. (eds.) Human Mental Workload, pp. 239–250. North Holland Press, Amsterdam (1988)

Hoffmann, J.: Vorhersage und Erkenntnis. Die Funktion von Antizipationen in der menschlichen Verhaltenssteuerung und Wahrnehmung. Hogrefe, Göttingen (1993)

Hoffmann, J.: Anticipatory behavioral control. In: Butz, M., Sigaud, O., Gerard, P. (eds.) Anticipatory Behavioral Control in Adaptive Learning Systems. Foundations, Theories and Systems, pp. 44–65. Springer, Berlin (2003)

Kallus, K. W.: Situationsbewusstsein und antizipative Prozesse. *Zeitschrift für Arbeitswissenschaft*, 63, 17–22 (2009)

Kallus, K.W., Barbarino, M., Van Damme, D.: Model of the Cognitive Aspects of Air Traffic Control (HUM.ET1.STO1.1000-REP-2). Eurocontrol, Brussels (1997)

Kallus, K.W., Tropper, K.: Anticipatory processes in critical flight situations. Proceedings to the 27th Conference of the European Association for Aviation Psychology, Potsdam, Germany, pp. 38–40 (2006)

Kallus, K.W., Tropper, K.: Anticipatory processes in critical flight situations. Proceedings of the 14th International Symposium on Aviation Psychology, Dayton, OH, 23–26 Apr, 2007

Koglbauer, I., Kallus K.W., Braunstingl, R., and Boucsein, W.: Recovery training in simulator improves performance and psychophysiological state of pilots during simulated and real VFR flight. *International Journal of Aviation Psychology*, 21 (2011, in print)

Pezzulo, G., Hoffmann, J., Falcone, R.: Anticipation and anticipatory behavior. Cogn. Process. **8**, 67–70 (2007)

Chapter 5
The Effects of Colored Light on Valence and Arousal

Investigating Responses Through Subjective Evaluations and Psycho-Physiological Measurements

Rosemarie J.E. Rajae-Joordens

Abstract Although red light is often said to be activating and blue light is thought to be relaxing, studies provide no unequivocal evidence for such a claim. There are indications that the effect on arousal evoked by colored light should not be attributed to its hue, but to uncontrolled variations in lightness and saturation instead. Moreover, cognitive processes, such as associations, also play a role. Therefore, not only arousal but also valence should be considered when studying the effect of colored light. In the current study, the effect of hue (red, green and blue), lightness and saturation of colored light on arousal and valence was investigated. Red light was found to be less pleasant and more arousing than green and blue light as measured by subjective evaluations, and as expected, saturated light was assessed to be more arousing than desaturated light. Conversely, no clear psycho-physiological effects were found. In conclusion, a discrepancy between questionnaires and psycho-physiological measurements has occurred in this study; its cause has not been identified yet.

5.1 Introduction

In earlier days, a number of researchers started to investigate the relation between colored light and arousal. Arousal is a physiological and psychological state involving the activation of the reticular activating system in the brain stem, the autonomic nervous system, and the endocrine system (Frijda, 1986). Changes in arousal affect the activity of the sympathetic nervous system, i.e. one of the two subsystems of the autonomic nervous system, which can be monitored by psycho-physiological parameters, such as skin conductivity and heart rate. Gerard (1958) found that exposure to red light increased arousal reflected by augmented systolic blood pressure,

R.J.E. Rajae-Joordens (✉)
Philips Research, Eindhoven, The Netherlands
e-mail: roos.rajae@philips.com

J. Westerink et al. (eds.), *Sensing Emotions*, Philips Research Book Series 12,
DOI 10.1007/978-90-481-3258-4_5, © Springer Science+Business Media B.V. 2011

skin conductance, respiration rate, eye blink frequency, cortical activation and subjective evaluations as compared to blue light. Wilson (1966) reported a different influence of red and green light on skin conductance with red light being more arousing than green light. In addition, Ali (1972) found more cortical activity after exposure to red light compared to blue light. Based on these findings, it was claimed that red light has the capacity to arouse and activate people, while blue and green light possess qualities that can calm individuals.

However, not all studies support this claim that red light is activating and blue light is calming. Thapan et al. (2001) demonstrated that the production of nocturnal melatonin, a pineal hormone that promotes sleepiness and lowers of the body temperature, is largely suppressed by blue light ($\lambda = 475$ nm), slightly by green light ($\lambda = 510$ nm), and hardly or not by red light ($\lambda = 650$ nm). This finding implies that blue light is most arousing, followed by green, while red light is least arousing.[1] Evidence for such a wavelength dependent effect might be provided by the following two studies. Nourse and Welch (1971) found that purple and blue light increased skin conductance and were therefore considered to be more stimulating than green light. Also, Lockley et al. (2006) demonstrated that blue light more effectively suppressed EEG delta and theta activity, i.e. indicators of respectively slow wave sleep and drowsiness, than green light, implying that blue light is more arousing than green light. In contrast with the studies mentioned above, however, Erwin et al. (1961) did not find any significant effect of red, yellow, and blue light on suppression of EEG alpha activity, i.e. indicator of relaxation at all. Considering the ambiguous effects of colored light on arousal discussed above, it is not clear yet whether relative short-term exposure of different colored illumination levels induce different psycho-physiological activity.

Although it is not known yet whether the effect of colored pigment, i.e. ink or paint, is comparable to the effect of colored light, it is worthwhile to take a closer look at studies examining the effect of pigment on arousal. Again, contradictory findings have been found. On one hand, blue or bluish-green color cards were found to be more arousing than red color card as expressed by lower EEG alpha and theta activity (Yoto et al., 2007) and in subjective evaluations (Valdez and Merhabian, 1994). On the other hand, strong wall colors, especially red, were shown to decrease EEG delta activity, putting the brain into a more excited state (Küller et al., 2008). Jacobs and Hustmyer (1974) studied the effect of colored cards on arousal; they found an arousing effect of red color cards on skin conductance and heart rate compared to yellow and blue color cards, but not on respiration. Mikellides (1990) also investigated the effect of colored walls on arousal, and found neither an effect on EEG, heart rate nor skin conductance. Suk (2006) did not find any effect of color cards on subjective evaluations either. In conclusion, the results on colored pigment do not help to clarify how colored light affects arousal.

[1]Note that in case suppression of melatonin production is the only mechanism behind the arousing effect of blue light, no arousing effect of blue light should be seen during daytime; melatonin is, after all, only produced at night, and can therefore not be suppressed during day time.

Interestingly, Kaiser (1984) and Robinson (2004) noticed that the green and red stimuli in early experiment of Wilson were not equated with regard to lightness and saturation levels (Wilson, 1966). They suggested that saturation differences might have been responsible for the effects instead of color differences. Also Mikellides (1990) concluded that it is not hue, but rather saturation that determines how exciting or calming a color is perceived. Higher saturation (Suk, 2006; Valdez and Mehrabian, 1994) and bright light (Cajochen, 2007; Kubota et al., 2002) are known to increase arousal levels. Thus, in order to get a better understanding of the way colored light affects arousal, a properly controlled study should be performed.

Not only an improper design, but also cognitive processes influence arousal. For example, Yoto et al. (2007) found that blue color elicited a higher arousal compared to red as expressed by lower EEG alpha and theta activity. Remarkably, in contrast with this psycho-physiological effect, the participants rated red to be more arousing than blue. Because red color was also found to more strongly activate the areas of perception and attention of the central cortical region in this study, the researchers suggested that blue is biologically activating, while red possibly elicits an anxiety state. As the color red is frequently used as a warning sign in dangerous situations (e.g. red traffic light, red stop signs, "code red" in alerting systems, red fire brigade trucks), we have learned to pay particular attention to the color red. This idea is supported by the observations of Gerard (1958) that red light evoked a variety of unpleasant associations related to blood, injuries, fire and danger, while blue light was associated with positive thoughts such as friendliness, romantic love and blue skies. Thus, increased arousal not necessarily means that one feels positively energized and active, but can instead be an indication of feelings of anger, fear or discomfort. As a consequence, not only arousal but also valence, i.e. the intrinsic attractiveness (positive valence) or aversiveness (negative valence), of a color should be taken into account when investigating the way a particular color affects people.

As mentioned before, a change in arousal affects the activity of the sympathetic nervous, which can be psycho-physiologically measured at a variety of places at the body. Some signals, however, are very small and difficult to capture. For example, EEG needs sensitive equipment that amplifies the signals as well as complex analyses compared to e.g. heart rate or respiration rate. Especially skin conductivity has been proven to be a relatively easy and reliable indicator of arousal. It has been used in a number of virtual reality studies in order to investigate feelings of excitement, immersion and presence (Lombard et al., 2000; Meehan et al., 2002; Rajae-Joordens, 2008; Rajae-Joordens et al., 2005; Wiederhold et al., 2001). No such clear objective psycho-physiological measurement for valence has been identified so far. Nowadays, the most common way to obtain data related to perceived pleasantness of a stimulus is by simply asking the involved person to communicate his experience. A disadvantage of this approach is, however, that commenting through introspection severely interrupts ongoing behavior. Therefore, it is highly desirable to find, next to the psycho-physiological correlates for arousal, a psycho-physiological equivalent for valence too.

In the current study, the short-term effect of hue (red, green and blue), lightness and saturation of colored light on arousal and valence was investigated. Twenty participants were exposed to 12 carefully defined light stimuli with equated lightness and saturation levels, and valence and arousal were investigated by means of subjective evaluations and a variety of objective psycho-physiological measurements, derived from skin conductance, heart rate, respiration, and skin temperature. Stimulus duration of 1 min was chosen similar to the experiments of Wilson (1966), Nourse and Welsch (1971), Jacobs and Hustmyer (1974), because substantially longer stimulus durations of e.g. 10 min might induce unwanted and uncontrolled side effects, such as boredom, annoyance and sleepiness. Because no psycho-physiological correlate for valence has been identified so far, an additional aim of this study was to derive one from the series of objective psycho-physiological measurements captured in this study.

5.2 Material and Method

5.2.1 Participants

In total, 20 Philips Research employees (8 females, 12 males) aged 22–55 years (mean $\pm$ S.D. $= 28 \pm 8$ years) without any form of color blindness participated in this experiment.

5.2.2 Experimental Setup

5.2.2.1 Light Stimuli

The experiment was performed in an empty test room with no incoming daylight and white painted walls and ceiling. Five light-emitting diode (LED) wall washers (RGB, 16 LEDs per color, DMX-driven) were located on the floor in such a way that their light output fully covered one of the walls. In order to investigate the effect of hue, saturation and lightness, for each of the 3 hues (red, green and blue), four wall washer settings with different saturation levels (2 levels: saturated vs. desaturated) and lightness levels (2 levels: light vs. dim), as well as a neutral setting for in between stimuli intervals were defined.

The colored light outputs of the wall washers were matched as closely as possible by means of 1976 CIELAB coordinates (L^*, a^* and b^*). The neutral setting obtained by means of 6 DMX-driven fluorescent light units integrated in the ceiling (4,300 K, 500 cd/m^2) was chosen as a reference. Lightness (L^*), saturation (defined as $(\sqrt{(a^*)^2 + (b^*)^2}/L^*)$) and hue (calculated via $(\tan^{-1}(b^*/a^*))$) were carefully controlled in order to allow comparisons of effects found. Measurements were taken with a photometer (Photo Research, Inc).

Due to technical limitations of the wall washers, the blue LEDs were substantially less powerful than the red and green LEDs. Lowering the lightness levels of the

Table 5.1 Overview of the mean Hue (H in degrees), Saturation (S) and Lightness (L^* in %) values of the 12 light stimuli

Hue	Red		Green		Blue	
Saturation	Saturated	Desaturated	Saturated	Desaturated	Saturated	Desaturated
L*=81%	H=21° S=2.3	H=21° S=0.8	H=162° S=2.3	H=162° S=0.8	– –	– –
L*=63%	H=21° S=2.3	H=21° S= 0.8	H=162° S=2.3	H=162° S=0.8	H=262° S=2.3	H=262° S=0.8
L*=54%	– –	– –	– –	– –	H=262° S=2.3	H=262° S=0.8

red and green LEDs to the maximal output level of the blue LEDs resulted in lightness levels that approximated their minimum. Consequently, further lowering these lightness levels of the red and the green LEDs to investigate lightness effects was impossible. In order to overcome this difficulty, an incomplete experimental design with a shift in lightness levels for the blue light stimuli was chosen. In Table 5.1, the hue, saturation and lightness values of the 12 light stimuli are summarized. Next, 12 scripts with different presentation orders for the 12 stimuli were prepared to control possible order effects.

5.2.2.2 Subjective Evaluations

By means of a questionnaire, participants were asked to indicate how they experienced the 12 light stimuli. For each light stimulus, participants were asked to write down their associations and to assess arousal (AROUS) and valence (VAL) on the valence dimension (ranging from very pleasant to very unpleasant) and arousal dimension (ranging from very activating to very calming) of the pictorial five-point scales of the Self-Assessment-Manikin (Lang, 1995). The end of the questionnaire concluded with the question which of the three stimulus colors, i.e. red, green and blue, the participant in general preferred most.

5.2.2.3 Objective Measurements

The NEXUS-10 (Mind Media BV, The Netherlands) was used to capture psycho-physiological responses triggered by the 12 light stimuli. Besides the most generally accepted measurement for arousal, i.e. skin conductivity, all further available sensors of the Nexus were also connected to gather a maximum of psycho-physiological data, because one of the aims of this study was to find a psycho-physiological correlate for valence too. Blood volume pulse, temperature and skin conductance were measured respectively by means of the blood volume pulse sensor clipped on the middle finger, the temperature sensor taped to the little finger, and two active skin conductance electrodes on the index and ring finger, all on the left hand. Respiration was measured by means of the respiration sensor belt put over the clothes around the chest. Eye movements were captured respectively by four passive electrodes attached on the face and a passive electrode on the neck as a reference, but due to time constraints and the unavailability of a suitable algorithm, this signal was not analyzed. All psycho-physiological data were stored by means of BioTrace+ Software, version 2008a (Mind Media BV, the Netherlands).

5.2.3 Experimental Procedure

Participants were invited to take place on a chair facing the white wall with the wall washers at a 3.5-m distance. Electrodes and other Nexus sensors were attached to the participant's body. After a short instruction, the test leader simultaneously started the BioTrace+ software to capture psycho-physiological measurements and one of the 12 scripts to drive the lights in the test room, and left the room to prevent any disturbance towards the participant.

Before presenting the first colored light stimulus, a baseline was recorded for 3 min in the neutral light setting. Subsequently, the fluorescent lights were turned off and 12 predefined colored light stimuli were presented 1 min each. An interval of 1 min was chosen between two colored light stimuli, in which the neutral light setting was set.

At the end of the experiment, the psycho-physiological measurements were stopped and all light stimuli were presented for a second time such that participants could see the light stimuli while filling out the questionnaire.

5.2.4 Data Processing

5.2.4.1 Subjective Evaluations

Due to the complexity of the data set, parametric multivariate analyses were highly preferred above non-parametric univariate analyses. Because the data of different participants were independent and the distance between points of the pictorial Likert-type questionnaire scales was assumed to be equal at all parts along that scale, these data were eligible for a parametric analysis on the condition that the data are normally distributed.

5.2.4.2 Psycho-Physiological Measurements

Raw psycho-physiological data, i.e. skin conductance (32 samples/s), temperature (32 samples/s), blood volume pressure (128 samples/s), and respiration (32 samples/s), were exported from the commercially available BioTrace+ Software, version 2008a, in order to be prepared for statistical analyses. By means of the Biosignal Toolbox (internally developed by Gert-Jan de Vries and Stijn de Waele, Philips Research, The Netherlands), the following 9 psycho-physiological measurements were derived from the exported data:

- Skin conductance level (SCL) in μSiemens;
- Number of skin conductance responses (#SCR);

- Skin temperature (ST) in °C;
- Skin temperature slope (ΔST);
- Heart rate (HR) in beats/min;
- Heart rate variability (HRV);
- Respiration depth (RD);
- Respiration rate (RR) in breaths/min;
- And coherence, i.e. correlation between respiration and heart rate (COH).

In short, the Biosignal Toolbox calculated HR, HRV, RD, RR and #SCR from respectively the blood volume pressure, respiration and skin conductance signals. The other four parameters, i.e. SCL, ST, ΔST and COH (computed by the BioTrace+ Software using the blood volume pressure and respiration values) did not need any further processing. Subsequently, by means of the Toolbox wrapper function, for each participant and for each of the 9 psycho-physiological parameters listed above, a mean for each of the 12 colored light stimuli presentation periods was calculated. The data obtained during baseline recording and in the neutral in between stimuli intervals were further omitted from the analyses.

5.3 Results

As mentioned earlier, due to technical limitations of the blue LEDs, it was not possible to investigate the effects of Hue, Saturation and Lightness on the subjective evaluations and objective measurements in a single analysis. Therefore, three separate analyses were defined (see Table 5.2), namely:

- A *RGB-analysis* to investigate the effect of Hue and Saturation over the *red*, *green* and *blue* 63%-Lightness stimuli;
- A *RG-analysis* to perform a complete analysis on Hue, Saturation and Lightness on *red* and *green* stimuli only;
- And a *B-analysis* to investigate the effect of Saturation and Lightness for the *blue* light stimuli.

Table 5.2 Visual representation of the stimuli used in the "Hue – Saturation" analysis (*RGB*), "Hue – Saturation – Lightness" analysis (*RG*), and the "Saturation – Lightness" analysis (*B*)

Hue	Red		Green		Blue	
Saturation	Saturated	Desaturated	Saturated	Desaturated	Saturated	Desaturated
81%-lightness	RG	RG	RG	RG	–	–
63%-lightness	RG/RGB	RG/RGB	RG/RGB	RG/RGB	RGB/B	RGB/B
54%-lightness	–	–	–	–	B	B

5.3.1 *Subjective Evaluations*

5.3.1.1 Data Simplification

Normality tests were performed on the data of the questionnaires first. These tests confirmed that the VAL ($p = 0.544$) and AROUS ($p = 0.914$) scores of the questionnaire showed a normal distribution, and therefore, it was proven to be legitimate to proceed with parametric tests. Next, possible order and gender effects had to be excluded. Therefore, multivariate ANOVAs with respectively VAL and AROUS as within-subject factor (12 levels: one for each light stimulus) and Order (2 levels: Started with saturated stimuli vs. Started with desaturated stimuli) or Gender (2 levels: male vs. female) as between-subject factor were executed. The results of these analyses are summarized in Table 5.3A. As can be seen, the factors Order and Gender did not affect VAL and AROUS, and can therefore be omitted from further analyses.

5.3.1.2 Mean Subjective Evaluations

Mean VAL and AROUS scores for all 12 light stimuli were calculated. The results are depicted in Fig. 5.1. In addition, mean VAL and AROUS (± S.E.M.) for the three hue levels, the two saturation levels, and the three lightness levels in the *RGB*-, *RG*- and *B-analysis* were calculated. These scores can be found in respectively Tables 5.4, 5.5, and 5.6.

5.3.1.3 Hue – Saturation (*RGB-Analysis*)

VAL and AROUS of the saturated and desaturated 63%-Lightness stimuli were analyzed by means of a repeated measurements ANOVA with Saturation (2 levels:

Table 5.3 Overview of the order and gender effects for the questionnaire scores and psychophysiological measurements (A), and the overall mean (± S.E.M.) for each of these measurements averaged over all 12 light stimuli (B)

A	Order effect	Gender effect	B	Mean ± S.E.M
VAL	F(12,7)=1.997; p=0.183	F(12,7)=0.630; p=0.770	VAL	3.400 ± 0.065
AROUS	F(12,7)=0.539; p=0.835	F(12,7)=1.034; p=0.504	AROUS	2.142 ± 0.069
SCL	F(12,7)=0.554; p=0.824	F(12,7)=0.982; p=0.535	SCL	5.575 ± 0.282
#SCR	F(12,7)=1.793; p=0.224	F(12,7)=1.499; p=0.303	#SCR	0.057 ± 0.004
ST	F(12,7)=0.750; p=0.685	F(12,7)=1.060; p=0.490	ST	32.14 ± 0.230
ΔST	F(12,7)=3.425; p=0.056	F(12,7)=0.319; p=0.960	ΔST	0.001 ± 0.001
HR	F(12,7)=0.678; p=0.678	F(12,7)=0.482; p=0.873	HR	72.34 ± 0.635
HRV	F(12,7)=2.625; p=0.104	F(12,7)=2.401; p=0.067	HRV	3.108 ± 0.215
RD	F(12,5)=5.629; p=0.034*	F(12,5)=0.803; p=0.652	RD	10.896 ± 0.564
RR	F(12,5)=1.094; p=0.496	F(12,5)=0.353; p=0.935	RR	14.957 ± 0.243
COH	F(12,7)=2.573; p=0.108	F(12,7)=4.007; p=0.037*	COH	–0.084 ± 0.016

*Significant at α=5%.

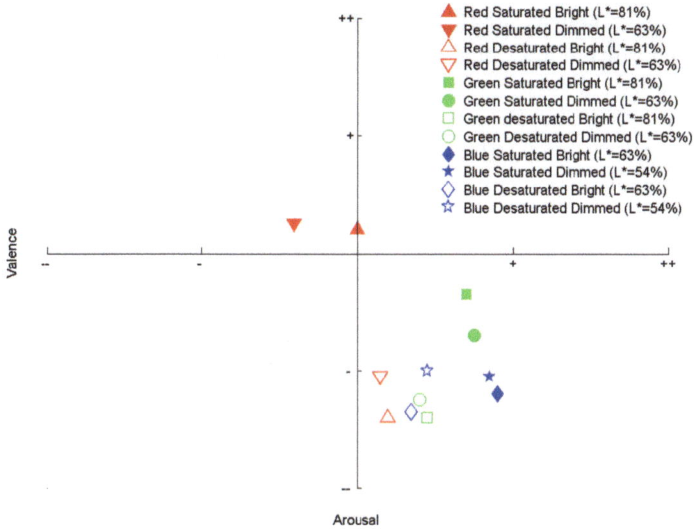

Fig. 5.1 Mean valence and arousal score on the questionnaire for each of the 12 light stimuli

Table 5.4 Mean VAL and AROUS (± S.E.M.) for the three hue levels in the *RGB*-, *RG*- and *B-analysis*

Hue	VAL_{RGB}	VAL_{RG}	VAL_B	$AROUS_{RGB}$	$AROUS_{RG}$	$AROUS_B$
Red	2.88 ± 0.15	2.99 ± 0.14	–	2.60 ± 0.17	2.50 ± 0.16	–
Green	3.58 ± 0.13	3.58 ± 0.11	–	2.03 ± 0.17	2.08 ± 0.15	–
Blue	3.63 ± 0.19	–	3.64 ± 0.11	1.73 ± 0.17	–	1.85 ± 0.11

Table 5.5 Mean VAL and AROUS (± S.E.M.) for the two saturation levels in the *RGB*-, *RG*- and *B-analysis*

Saturation	VAL_{RGB}	VAL_{RG}	VAL_B	$AROUS_{RGB}$	$AROUS_{RG}$	$AROUS_B$
Saturated	3.41 ± 0.13	3.26 ± 0.16	3.88 ± 0.16	2.45 ± 0.16	2.85 ± 0.20	1.88 ± 0.15
Desaturated	3.30 ± 0.16	3.30 ± 0.15	3.40 ± 0.24	1.78 ± 0.14	1.73 ± 0.11	1.83 ± 0.23

Table 5.6 Mean VAL and AROUS (± S.E.M.) for the three lightness levels in the *RGB*-, *RG*- and *B-analysis*

Lightness	VAL_{RGB}	VAL_{RG}	VAL_B	$AROUS_{RGB}$	$AROUS_{RG}$	$AROUS_B$
81%	–	3.34 ± 0.11	–	–	2.26 ± 0.15	–
63%	–	3.23 ± 0.12	3.63 ± 0.19	–	2.31 ± 0.15	1.73 ± 0.17
54%	–	–	3.65 ± 0.18	–	–	1.98 ± 0.18

saturated vs. desaturated) and Hue (3 levels: red vs. green vs. blue) as within-subject factors. The results of these analyses are summarized in Tables 5.7 and 5.8.

As can be seen in Table 5.7, a significant effect of Saturation on AROUS, but not on VAL was found. The desaturated stimuli were judged significantly less arousing than the saturated stimuli, while they were equally positively rated with regard to valence (see Table 5.4).

Hue, on the other hand, significantly influenced both VAL and AROUS. The red stimuli were scored significantly less pleasant and more arousing than the green (VAL: $p = 0.001$; AROUS: $p = 0.008$) and blue (VAL: $p = 0.021$; AROUS: $p < 0.001$) stimuli as revealed by post-hoc tests. The latter two were equally rated with regard to VAL ($p = 1.00$) and AROUS ($p = 0.207$). In Table 5.5, the mean VAL and AROUS scores per hue can be found.

Table 5.8 shows a significant interaction effect between Saturation and Hue for VAL and AROUS. The interaction effect on VAL was due to the fact that the saturated blue stimulus was rated significantly more pleasant than the desaturated blue stimulus (Δ[2] $= 0.55 \pm 0.21$; $p = 0.017$), while for the red stimuli the opposite was true. The saturated red stimulus was rated, although just not significant, less pleasant than the desaturated red stimulus ($\Delta = -0.55 \pm 0.28$; $p = 0.061$). The saturated and desaturated green stimuli were scored equally pleasant ($\Delta = 0.35 \pm 0.28$; $p = 0.232$). Further, the interaction effect on AROUS was caused by the fact that the difference between saturated and desaturated stimuli was very significant for red ($\Delta = 1.30 \pm 0.27$; $p < 0.001$) and green ($\Delta = 0.55 \pm 0.11$; $p < 0.001$), and not significant for blue ($\Delta = 0.15 \pm 0.21$; $p = 0.481$).

5.3.1.4 Hue – Saturation – Lightness (*RG-Analysis*)

VAL and AROUS were analyzed by means of a repeated measurements ANOVA with Saturation (2 levels: saturated vs. desaturated), Lightness (2 levels: dimmed vs. bright) and Hue (2 levels: red vs. green) as within-subject factors. In Tables 5.7 and 5.8 an overview of the results is given.

Similar to the *RGB-analysis*, Saturation was found to significantly affect AROUS, while no effect on VAL was found. The saturated stimuli were judged significantly more arousing than the desaturated stimuli, while they were equally positively rated with regard to valence (see Table 5.4).

Moreover, also in line with the *RGB-analysis*, both VAL and AROUS were significantly influenced by Hue. The red light stimuli again scored significantly lower on VAL and higher on AROUS than the green stimuli. In Table 5.5, the mean VAL and AROUS scores per hue are listed.

In this analysis, three interaction effects were found. Again, as in the *RGB-analysis*, the interaction between Saturation and Hue was significant for VAL and AROUS. Saturated green stimuli were rated, although not significant, more pleasant as compared to the desaturated green ($\Delta = 0.30 \pm 0.28$; $p = 0.289$), while saturated

[2]The symbol "Δ" is used to indicate the "mean difference ± S.E.M.".

Table 5.7 Overview of the main effects of saturation, lightness and hue in the *RGB-analysis*, *RG-analysis* and *B-analysis*

Main	Saturation			Lightness			Hue		
measure	RGB	RG	B	RGB	RG	B	RGB	RG	B
VAL	F(1,19)=0.437; p=0.517	F(1,19)=0.056; p=0.870	F(1,19)=5.818; p=0.026*	–	F(1,19)=1.879; p=0.186	F(1,19)=0.056; p=0.815	F(2,18)=10.217; p=0.001*	F(1,19)=18.983; p<0.001*	–
AROUS	F(2,18)=31.667; p<0.001*	F(1,19)=43.475; p<0.001*	F(1,19)=0.054; p=0.818	–	F(1,19)=0.248; p=0.624	F(1,19)=4.524; p=0.047*	F(2,18)=10.866; p=0.001*	F(1,19)=13.693; p=0.002*	–
SCL	F(1,19)=0.172; p=0.683	F(1,19)=0.358; p=0.566	F(1,19)=0.830; p=0.374	–	F(1,19)=0.223; p=0.642	F(1,19)=0.050; p=0.826	F(2,18)=1.710; p=0.209	F(1,19)=0.410; p=0.530	–
#SCR	F(1,19)=0.294; p=0.594	F(1,19)=0.167; p=0.687	F(1,19)=0.340; p=0.567	–	F(1,19)=0.024; p=0.879	F(1,19)=0.948; p=0.342	F(2,18)=0.765; p=0.480	F(1,19)=0.077; p=0.785	–
ST	F(1,19)=1.340; p=0.261	F(1,19)=1.082; p=0.311	F(1,19)=0.522; p=0.479	–	F(1,19)=1.105; p=0.306	F(1,19)=0.027; p=0.871	F(2,18)=0.318; p=0.732	F(1,19)=0.381; p=0.544	–
ΔST	F(1,19)=0.857; p=0.366	F(1,19)=0.218; p=0.646	F(1,19)=0.859; p=0.366	–	F(1,19)=0.144; p=0.709	F(1,19)=0.027; p=0.870	F(2,18)=1.233; p=0.315	F(1,19)=0.070; p=0.794	–
HR	F(1,19)=0.436; p=0.517	F(1,19)=0.781; p=0.388	F(1,19)=0.744; p=0.399	–	F(1,19)=0.048; p=0.829	F(1,19)=5.329; p=0.032*	F(2,18)=1.080; p=0.361	F(1,19)=0.070; p=0.795	–
HRV	F(1,19)=4.130; p=0.056	F(1,19)=2.700; p=0.117	F(1,19)=1.450; p=0.243	–	F(1,19)=2.356; p=0.141	F(1,19)=2.932; p=0.103	F(2,18)=0.055; p=0.946	F(1,19)=0.486; p=0.494	–
RD	F(1,17)=0.003; p=0.957	F(1,17)=0.151; p=0.703	F(1,19)=0.160; p=0.694	–	F(1,17)=0.432; p=0.520	F(1,19)=0.109; p=0.745	F(2,16)=0.150; p=0.861	F(1,17)=0.168; p=0.687	–
RR	F(1,17)=1.540; p=0.231	F(1,17)=1.567; p=0.228	F(1,19)=2.705; p=0.116	–	F(1,17)=0.220; p=0.645	F(1,19)=0.082; p=0.777	F(2,16)=1.175; p=0.334	F(1,17)=2.350; p=0.144	–
COH	F(1,18)=4.959; p=0.039*	F(1,18)=6.091; p=0.024*	F(1,18)=2.302; p=0.147	–	F(1,18)=0.384; p=0.543	F(1,18)=0.639; p=0.434	F(2,17)=0.556; p=0.583	F(1,18)=3.297; p=0.086	–

*Significant at α=5%.

Table 5.8 Overview of the two-way interaction effects between saturation, lightness and hue in the *RGB-analysis*, *RG-analysis* and *B-analysis*

Interaction measure	Saturation X Lightness RGB	RG	B	Saturation X Hue RGB	RG	B	Lightness X Hue RGB	RG	B
VAL	–	F(1,19)=0.192; p=0.666	F(1,19)=1.000; p=0.330	F(2,18)=6.293; p=0.008*	F(1,19)=6.153; p=0.023*	–	–	F(1,19)=1.572; p=0.225	–
AROUS	–	F(1,19)=8.329; p=0.009*	F(1,19)=0.884; p=0.359	F(2,18)=4.394; p=0.028*	F(1,19)=12.30; p=0.002*	–	–	F(1,19)=1.331; p=0.263	–
SCL	–	F(1,19)=1.457; p=0.242	F(1,19)=0.286; p=0.599	F(2,18)=0.765; p=0.480	F(1,19)=0.014; p=0.908	–	–	F(1,19)=0.661; p=0.426	–
#SCR	–	F(1,19)=0.957; p=0.340	F(1,19)=2.110; p=0.163	F(2,18)=3.422; p=0.055	F(1,19)=3.580; p=0.074	–	–	F(1,19)=0.980; p=0.335	–
ST	–	F(1,19)=0.002; p=0.962	F(1,19)=0.437; p=0.517	F(2,18)=0.121; p=0.886	F(1,19)=0.216; p=0.647	–	–	F(1,19)=0.007; p=0.934	–
ΔST	–	F(1,19)=0.991; p=0.332	F(1,19)=1.569; p=0.226	F(2,18)=0.744; p=0.489	F(1,19)=0.904; p=0.354	–	–	F(1,19)=3.909; p=0.063	–
HR	–	F(1,19)=0.200; p=0.660	F(1,19)=0.710; p=0.410	F(2,18)=1.251; p=0.310	F(1,19)=0.479; p=0.497	–	–	F(1,19)=0.560; p=0.463	–
HRV	–	F(1,19)=0.489; p=0.493	F(1,19)=0.257; p=0.618	F(2,18)=0.972; p=0.398	F(1,19)=0.708; p=0.410	–	–	F(1,19)=0.444; p=0.513	–
RD	–	F(1,17)=4.135; p=0.058	F(1,19)=2.872; p=0.106	F(2,16)=1.484; p=0.256	F(1,17)=0.232; p=0.636	–	–	F(1,17)=0.012; p=0.914	–
RR	–	F(1,17)=0.063; p=0.804	F(1,19)=1.548; p=0.229	F(2,16)=0.085; p=0.919	F(1,17)=0.000; p=0.995	–	–	F(1,17)=0.097; p=0.759	–
COH	–	F(1,18)=1.859; p=0.190	F(1,18)=0.066; p=0.801	F(2,17)=1.698; p=0.213	F(1,18)=0.585; p=0.454	–	–	F(1,18)=0.032; p=0.859	–

*Significant at α=5%.

red were rated, although not significant, more unpleasant than the desaturated red stimuli ($\Delta = -0.38 \pm 0.25$; $p = 0.152$). Also, the difference in arousal between saturated and desaturated stimuli was very significant for both red ($\Delta = 1.45 \pm 0.23$; $p < 0.001$) and green ($\Delta = 0.80 \pm 0.15$; $p < 0.001$) hues.

Additional to the findings of the *RGB-analysis*, a significant interaction between Saturation and Lightness was found on AROUS. The difference in arousal between bright saturated and bright desaturated stimuli was very significant ($\Delta = 1.33 \pm 0.21$; $p < 0.001$), but less significant for the dimmed saturated and dimmed desaturated stimuli ($\Delta = 0.93 \pm 0.16$; $p < 0.001$).

5.3.1.5 Saturation – Lightness (*B-Analysis*)

VAL and AROUS were analyzed by means of a repeated measurements ANOVA with Saturation (2 levels: saturated vs. desaturated) and Lightness (2 levels: dimmed vs. bright) as within-subject factors. In Tables 5.7 and 5.8, the results are summarized.

For VAL, only a significant effect of Saturation was found due to the fact that the saturated blue light stimuli were rated significantly more pleasant than the desaturated ones on the valance scale. This finding is deviant from the *RGB-* and *RG-analyses,* in which Saturation effects for VAL were not found.

With regard to AROUS, also only one significant effect was seen. AROUS was significantly affected by Lightness, however, in an unexpected way. The blue dimmed (54%-Lightness) light stimuli were rated to be significantly more arousing (see Table 5.6) than the blue bright (63%-Lightness) light stimuli; this finding was not seen in the *RG-analysis*.

5.3.1.6 Color Preference

At the end of the questionnaire, participants had to indicate which of the three stimulus colors, i.e. red, green and blue, they preferred overall. A majority of 50% of the participants preferred the blue stimuli, 40% had a preference for the green stimuli, and only 10% of the participants indicated the red stimuli to be their favorite. As expected, the preference scores correspond with the valence scores in this study; the highly preferred blue and green stimuli showed high VAL scores, while the low preference for the red light stimuli was reflected by low VAL scores.

5.3.2 Objective Measurements

5.3.2.1 Data Simplification

Similar to the analysis of the subjective evaluations, possible order and gender effects had to be excluded first. Therefore, multivariate ANOVAs with one of the 9 psycho-physiological measurements as within-subject factor (12 levels: one for each light stimulus) and Order (2 levels: Started with saturated stimuli vs. Started with desaturated stimuli) or Gender as between-subject factor (2 levels: male vs. female) were executed. In Table 5.3A, a summary of the results of these analyses is given. As can be seen, only a significant Order effect for RD and a significant Gender effect for COH were found. The Order effect for RD, however, completely disappeared when looking at the 12 separate between-subjects tests for each light stimulus, and therefore, the multivariate Order effect for RD was considered to be not meaningful. As a consequence, the factors Gender and Order were further omitted from analyses with exception of the analysis of COH, in which Gender will be taken into account.

5.3.2.2 Mean Psycho-Physiological Measurements

Mean SCL, #SCR, ST, ΔST, HR, HRV, RD, RR and COH levels for all 12 light stimuli, as well as one overall mean, were calculated. The overall means (± S.E.M.) for SCL, #SCR, ST, ΔST, HR, HRV, RD, RR and COH averaged over all 12 light stimuli can be found in Table 5.3B.

5.3.2.3 Hue – Saturation (*RGB-Analysis*)

Psycho-physiological measurements of the saturated and desaturated 63%-Lightness stimuli were analyzed by means of a repeated measurements ANOVAs with Saturation (2 levels: saturated vs. desaturated) and Hue (3 levels: red vs. green vs. blue) as within-subject factors. In the analysis of COH, an additional between-subject factor Gender (2 levels: male vs. female) was added. Results of the analyses are summarized in Tables 5.7 and 5.8.

Only two significant effects originating from the COH analysis were found. The significant main effect of Gender [$F(1,18) = 5.166$; $p = 0.036$] indicated a linear shift in COH, i.e. males (*m*[3] = –0.157 ± 0.043) showed lower COH levels over all light stimuli as compared to females ($m = 0.050 \pm 0.049$). As no significant interaction effect between Gender and Saturation [$F(1,18) = 1.010$; n.s.], and Gender and Hue [$F(2,17) = 0.148$; n.s.] were found, it can be concluded that males and females respond in a similar way to the light stimuli. Therefore, no further attention will be paid to this gender difference for COH.

The only relevant statistical significant effect was a main effect of Saturation (see Table 5.7) caused by the fact that COH levels for saturated stimuli ($m = -0.090 \pm 0.042$) were significantly lower than those for desaturated ones ($m = -0.017 \pm 0.054$).

[3]The term "*m*" is used to indicate "mean ± S.E.M.".

5.3.2.4 Hue – Saturation – Lightness (*RG-Analysis*)

Psycho-physiological measurements of the saturated and desaturated red and green light stimuli with 63%-lightness (dimmed) and 81%-lightness (bright) were analyzed by means of a repeated measurements ANOVA with Saturation (2 levels: saturated vs. desaturated), Lightness (2 levels: dimmed vs. light) and Hue (2 levels: red vs. green) as within-subject factors. In the analysis of COH, an additional between-subject factor Gender (2 levels: male vs. female) was added. See Tables 5.7 and 5.8 for an overview of the results.

Similar effects as in the *RGB-analysis* were found. Again, a significant effect of Gender was seen [$F(1,18) = 5.450$; $p = 0.031$]. Males ($m = -0.157 \pm 0.042$) showed significantly lower COH levels over all light stimuli than females ($m = 0.034 \pm 0.078$). Further, the interaction effect between Gender and Hue [$F(1,18) = 0.057$; n.s.] was again not significant. Finally, a main effect of Saturation was found again (see Table 5.7). The COH levels for saturated stimuli ($m = -0.108 \pm 0.037$) were also in this analysis significantly lower than those for desaturated ones ($m = -0.052 \pm 0.055$).

Additional to the *RGB-analysis*, the interaction between Gender and Lightness [$F(1,18) = 0.037$; n.s.] was tested, and was found to be not significant. This time, however, a significant interaction between Gender and Saturation for COH [$F(1,18) = 4.567$; $p = 0.047$] was seen. The difference between desaturated and saturated stimuli for COH levels was very significant for females ($\Delta = 0.127 \pm 0.042$; $p = 0.019$), but not significant for males ($\Delta = 0.009 \pm 0.035$; $p = 0.801$). This interaction effect, however, has not been seen in the *RGB-* and *B-analyses* and might probably be caused to coincidence. When ignoring this interaction effect, only a significant main effect of gender for COH remains. In other words, the COH levels of males and females respond in a similar way to the light stimuli, and therefore, the gender difference for COH will be further ignored.

5.3.2.5 Saturation – Lightness (*B-Analysis*)

Psycho-physiological measurements of the saturated and desaturated blue light stimuli with 54%-Lightness (dimmed) and 63%-Lightness (bright) were analyzed by means of a repeated measurements ANOVA with Saturation (2 levels: saturated vs. desaturated), and Lightness (2 levels: dimmed vs. light) as within-subject factors. In the analysis of COH, an additional between-subject factor Gender (2 levels: male vs. female) was added. An overview of the results of the analyses can be found in Tables 5.7 and 5.8.

In contrast with the *RGB-* and *RG-analyses*, only a Gender effect [$F(1,18) = 7.067$; $p = 0.016$], but no Saturation effect (see Table 5.7) on COH was found. Again, males ($m = -0.180 \pm 0.042$) showed significantly lower COH levels over all light stimuli than females ($m = 0.043 \pm 0.082$). As no significant interaction effect between Gender and Saturation [$F(1,18) = 0.515$; n.s.] and Gender and Lightness [$F(1,18) = 0.023$; n.s.] were found, it can again be concluded that males and females respond similar to the light stimuli.

Surprisingly, on the other hand, a significant main effect of Lightness was found on HR (see Table 5.7). HR was significantly higher for dimmed stimuli ($m = 73.13 \pm 2.08$) as compared to the bright light stimuli ($m = 71.42 \pm 2.36$).

5.4 Discussion

In the current study, the short-term effect of hue, lightness and saturation of colored light on arousal and valence was investigated. Twenty participants were exposed to 12 carefully defined red, green and blue light stimuli. Valence and arousal were investigated by means of objective psycho-physiological measurements and subjective evaluations. The results of the subjective evaluations show a number of significant effects, including very significant hue effects, while only a very few effects of psycho-physiological measurements were found.

5.4.1 Subjective Evaluations

5.4.1.1 Effect of Saturation

In line with previous studies (Suk, 2006; Valdez and Mehrabian, 1994), the desaturated stimuli were judged significantly less arousing than the saturated stimuli. This effect was found for the red and green stimuli only, and not for the blue light stimuli, as the saturation effect was strongest in the *RG-analysis*, less pronounced in the *RGB-analysis*, and not present in the *B-analysis* (see Tables 5.5 and 5.7). The absence of a saturation effect for the blue light stimuli might be explained by the fact that the desaturated blue light stimuli were rated to be unpleasant compared to the saturated blue light stimuli, probably due to their rather grayish instead of pale bluish appearance. No such saturation difference on pleasantness was found for the green and red light stimuli as the saturated and desaturated stimuli were rated equally pleasant. As already discussed in the introduction, unpleasant feelings might also increase arousal. Thus, it might well be the case that the arousing effect of the saturated blue light stimuli is masked by an unexpected arousing effect of the unpleasant grayish desaturated stimuli.

5.4.1.2 Effect of Lightness

In contrast with earlier studies showing an arousing effect of bright light (Cajochen, 2007; Kubota et al., 2002), no such effects were found in this study. The difference in lightness between the dimmed and bright red and green stimuli might have simply not been large enough to induce a lightness effect on arousal. Red and green bright and dimmed stimuli were found to be equally arousing. Surprisingly, bright blue stimuli were judged to be less arousing than the blue dimmed ones. Because the difference in lightness between the bright and dimmed blue stimuli was even smaller ($63\%–54\% = 9\%$) than the difference in lightness between the bright and dimmed red and green stimuli (81%–63% = 18%), it is highly unlikely that this unexpected reversed effect for the blue stimuli is a lightness effect. In conclusion, possibly due to a too small difference in lightness between the bright and dimmed light stimuli in this study, no clear effects of lightness on arousal have been found.

5.4.1.3 Effect of Hue

Red light stimuli were found to be significantly less pleasant and more arousing than the green and blue stimuli, while the latter two were judged to be equally pleasant and arousing. The observation of the unpleasant arousingness of the red light stimuli is in line with the findings of Jacob and Suess (1975), who reported higher anxiety scores, and therefore higher arousal levels, after a 15-min exposure to red slides as compared to green and blue slides. Earlier, Wexner (1954) also found that the color orange was associated with feelings of excitement, distress and disturbance, while the color green was associated with comfort, calmness and serenity. Kaya and Epps (2004) recently reported again that the color green evoked mainly positive emotions of relaxation and comfort. Thus, the findings of this study that the red light stimuli were experienced to be less pleasant and more arousing than the green and blue stimuli correspond with the findings of a number of earlier studies.

The difference in arousal between saturated and desaturated light stimuli was dependent of hue. This difference was very significant for the red, slightly significant for the green, and not significant for the blue light stimuli. These results suggest that the significant hue effects found, and possibly to a certain extent also the saturation effects found in this study, might have been caused by the unpleasantness and arousingness of the saturated red stimuli mainly. This idea is supported by the fact that the saturated red light stimuli were found unpleasant as compared to the desaturated red ones, while for the green and blue light stimuli the opposite was true.

5.4.2 Objective Measurements

With regard to the psycho-physiological data, coherence levels for saturated light stimuli were found to be significantly lower than those for desaturated light stimuli. Identically to the saturation effect found on the questionnaires, this effect was only found for the red and green, and not for the blue light stimuli. The saturation effect was strongest in the *RG-analysis*, less pronounced in the *RGB-analysis*, and not present in the *B-analysis*. In addition, against expectations, heart rate was found to be significantly higher for dimmed stimuli as compared to the bright light stimuli for the blue light stimuli only. Increased heart rate is considered to be an indication of increased arousal (Frijda, 1986). Interestingly, this lightness effect of heart rate coincided with the saturation effect found on the arousal questionnaire in the *B-analysis*. Based on these findings, it might be concluded that saturation and lightness affected arousal as measured by both objective psycho-physiological measurements and subjective evaluations in this study.

It should be noted, however, the mean coherence values were located around zero, and should be considered as very weak correlations. Moreover, the unexpected reversed lightness effect on heart rate was only found for the blue stimuli with a 9% difference in lightness between the bright and dimmed light stimuli, and not for the red or green light stimuli with a 18% difference in lightness between bright and dimmed light stimuli. Therefore, it is highly unlikely that this unexpected reversed effect for the blue stimuli is a lightness effect. Possibly, the unpleasantness of the rather grayish blue dimmed desaturated stimulus might have been responsible for a higher arousal level reflected by an increased heart rate. More research is needed to clarify this finding. Even more important, however, is the fact that the skin conductance measurements (#SCR and SCL), which are considered to be relatively strong indicators of arousal, show no significant effect. The fact that no single effect has been found on skin conductance can be seen as a strong argument to further ignore the very small effects of coherence and the single effect on heart rate. Thus, it can be concluded that hue, saturation and lightness did not differentially affect the psycho-physiological measurements in this study; consequently, the aim to find a psycho-physiological correlate for valence could not be reached.

Because it was not clear on forehand whether or not a hue effect should be expected, due to the ambiguous results reported in earlier studies, the absence of a hue effect on psycho-physiological measurements is not fully surprising. Already in early days, Erwin et al. (1961) reported similar results as found in the present study. Red, yellow, and blue light did not affect suppression of EEG alpha activity. Moreover, a closer look at the results of Wilson (1966), revealed no color differences with regard to the absolute skin conductance levels when exposing his participants to five pairs of red and green light stimuli either. Wilson only found a color effect on the maximum increase in skin conductance (SCR), from the level at the time of stimulus onset, occurring within the first 12 s of the 1 min exposure time. This effect, however, was established by a non-parametric sign test, and not via a univariate analysis of the effect size; the number of pairs in which the red light-induced SCR exceeded the green light-induced SCR was significantly higher than the number of pairs in which the green light-induced SCR exceeded the red light-induced SCR. Based on this result, Wilson concluded that red was more arousing than green. Because very small indifferent effects easily become significant in such a simple statistical test, however, the validity of Wilson's conclusion should be questioned. In conclusion, the absence of an obvious effect of colored light on the psycho-physiological measurements in this study is not an entirely new observation; instead, it has been seen in at least a few other studies before.

5.4.3 Subjective Evaluations Versus Objective Measurements

Nevertheless, combining the psycho-physiological results and subjective evaluations reveals a discrepancy between these two types of measurements. The question arises how this discrepancy can be explained. The fact that only effects were found on subjective evaluations and not on psycho-physiological measurements suggests that participants might have filled in the questionnaire on basis of experience, history and preference without being physically affected by the colored light stimuli. However, it might also be the case that the differences between the psycho-physiological responses triggered by the different colored light stimuli were simply very small, and therefore undetectable by the equipment used. Another explanation might be that the effects of colored light are only gradually evolving, implying that measuring periods of 1 min are too short to detect differences. Further research is needed to investigate these hypotheses.

5.5 Conclusion

In conclusion, hue (red, green and blue) and saturation of colored light have been found to affect arousal and valence as captured by subjective evaluations. No clear effects of lightness on the questionnaires were found. Psycho-physiological measurements, on the other hand, were not differentially affected at all by the different light stimuli. So far, it is not clear how this discrepancy between questionnaires and psycho-physiological can be explained.

Acknowledgments The author wants to thank Iris Hanique from the Radboud University Nijmegen, The Netherlands, for her excellent assistance in preparing and performing the experiment, and Gert-Jan de Vries for his excellent support in analyzing the psycho-physiological data. The author also wants to thank Dragan Sekulovski and Ingrid Vogels for their help with selecting the proper light stimuli for this study.

References

Ali, M.R.: Pattern of EEG recovery under photic stimulation by light of different colors. Electroencephalogr. Clin. Neurophysiol. **33**, 332–35 (1972)
Cajochen, C.: Alerting effects of light. Sleep Med. Rev. **11**, 453–464 (2007)
Erwin, C.W., Lerner, M., Wilson, N.J., Wilson, W.P.: Some further observations on the photically elicited arousal response. Electroencephalogr. Clin. Neurophysiol. **13**, 391–394 (1961)
Frijda, N.: The Emotions, pp. 138–190. Maison des Sciences de l'Homme and Cambridge University Press, Cambridge (1986)
Gerard, R.M.: Differential Effects of Coloured Lights on Psycho-Physiological Functions. Ph.D. Dissertation, University of California, Los Angeles (1958)
Jacob, K.W., Suess, J.F.: Effects of four psychological primary colors on anxiety state. Percept. Mot. Skills **41**, 207–210 (1975)
Jacobs, K.W., Hustmeyer, F.E.: Effects of four psychological primary colors on GSR, heart rate, and respiration rate. Percept. Mot. Skills **38**, 763–766 (1974)
Kaiser, P.: Physiological response to color: a critical review. Color Res. Appl. **9**, 29–36 (1984)
Kaya, N., Epps, H.: Relationship between color and emotion: a study of college students. Coll. Stud. J. **38**, 396–405 (2004)
Kubota, T., Uchiyama, M., Suzuki, H., Shibui, K., Kim, K., Tan, X., et al.: Effects of noctural bright light on saliva melatonin, core body temperature and sleep propensity rhythms in human subjects. Neurosci. Res. **42**, 115–122 (2002)
Küller, B., Mikellides, B., Janssens, J.: Color, arousal, and performance: a comparison of three experiments. Color Res. Appl. **34**, 141–152 (2008)

Lang, P.: The emotion probe. Am. Psychol. **50**, 372–385 (1995)
Lockley, S.W., Evans, E.E., Scheer, F.A., Brainard, G.C., Czeisler, C.A., Aeschbach, D.: Short-wavelength sensitivity for the direct effects of light on alertness, vigilance, and the waking electroencephalogram in humans. Sleep **29**, 161–168 (2006)
Lombard, M., Reich, R.D., Grabe, M.E., Bracken, C.C., Ditton, T.B.: Presence and television: the role of screen size. Hum. Commun. Res. **26**, 75–98 (2000)
Meehan, M., Insko, B., Whitton, M., Brooks, F.P.: Physiological measures of presence in stressful virtual environments. ACM Trans. Graph. **21**, 645–652 (2002)
Mikellides, B.: Color and psycho-physiological arousal. J. Architect. Plann. Res. **7**, 13–20 (1990)
Nourse, E.W., Welch, R.B.: Emotional attributes of colour: a comparison of violet and green. Percept. Mot. Skills **32**, 403–406 (1971)
Rajae-Joordens, R.J.E.: Measuring experiences in gaming and TV applications – investigating the added value of a multi-view autostereoscopic 3D display. In: Westerink, J., Ouwerkerk, M., de Ruyter, B., Pasveer, F., Overbeek, T. (eds.) Philips Research Series: Probing Experience – From Academic Research to Commercial Propositions, pp. 77–90. Springer, The Netherlands (2008)
Rajae-Joordens, R., Langendijk, E., Wilinski, P., Heynderickx, I. Added value of a multi-view autostereoscopic 3D display in gaming applications. Society for Information Display, Proceedings of the 12th International Display Workshops, Takamatsu, Japan, pp. 1731–1734 (2005)
Robinson, W.S.: Colors, arousal, functionalism, and individual differences. Psyche **10**, 2 (2004)
Suk, H.J.: Color and emotion. A study on the affective judgment across media and in relation to visual stimuli. Ph.D. Dissertation, University of Mannheim, Germany (2006)
Thapan, K., Arendt, J., Skene, D.: An action spectrum for melatonin suppression: Evidence for a novel non-rod, non-cone photoreceptor system in humans. J. Physiol. **535**, 261–267 (2001)
Valdez, P., Mehrabian, A.: Effects of colour on emotions. J. Exp. Psychol. Gen. **123**, 394–409 (1994)
Wexner, L.B.: The degree to which colors (Hues) are associated with moodtones. J. Appl. Psychol. **38**, 432–435 (1954)
Wiederhold, B.K., Dong, P.J., Kaneda, M., Cabral, I., Lurie, Y., May, T., et al.: An investigation into physiological responses in virtual environments: an objective measurement of presence. In: Riva, G., Calimberti, C. (eds.) Toward Cyberpsychology: Mind, Cognition and Society in the Internet Age. IOS Press, Amsterdam, 175–183 (2001)
Wilson, G.D.: Arousal properties of red versus green. Percept. Mot. Skills **23**, 947–949 (1966)
Yoto, A., Katsuura, T., Iwanga, K., Shimomura, Y.: Effects of object colour stimuli on human brain activities in perception and attention referred to EEG alpha band response. J. Physiol. Anthropol. **26**, 373–379 (2007)

Chapter 6
Audiovisual Expression of Emotions in Communication

Emiel Krahmer and Marc Swerts

Abstract Non-verbal cues may reveal a lot about the emotional state of a user. However, the way these expressions of emotion are often studied in the scientific literature may be rather different from the actual expressions in reality, which are dynamic, spontaneous and potentially multimodal. In this chapter, we systematically compare posed and spontaneous emotional expressions, which were collected using an experimental language-based induction method in which participants were asked to produce sentences with an increasingly emotional content. It was found that spontaneous positive and negative expressions lead to more positive and negative self-reported emotion scores than the posed ones. Interestingly, however, perception studies revealed that judges rate the posed dynamic facial expressions as significantly stronger than the spontaneous ones. Finally, it was studied whether better acting skills lead to more realistic expressions, which turned out not to be the case.

6.1 Introduction

For many practical applications, it would be helpful if the user's emotional state could be automatically recognized and responded to (e.g., Picard and Klein, 2002). Some researchers argue that tracking user emotions is best done with physiological measurements such as heart rate, blood pressure, skin conductance or cortisol levels (e.g., Picard et al., 2001, among many others), and while recent years have seen a lot of progress in our understanding of such measurements and their relation to affective state, they have disadvantages as well. For example, these measures may be difficult to interpret and the measuring can be intrusive and difficult to apply in practical applications. In this chapter, we want to explore an alternative possibility, namely paying attention to the non-verbal behavior of users.

In our interactions with others, such non-verbal behaviors play an important role. Most people find it relatively easy to determine a persons emotional state on the basis of his intonation, gestures and facial expressions. There is a large body of

E. Krahmer (✉)
Faculty of Humanities, Tilburg centre for Cognition and Communication (TiCC), Tilburg University, Tilburg, The Netherlands
e-mail: E.J.Krahmer@uvt.nl

J. Westerink et al. (eds.), *Sensing Emotions*, Philips Research Book Series 12,
DOI 10.1007/978-90-481-3258-4_6, © Springer Science+Business Media B.V. 2011

scientific research to back up this claim, especially where facial expressions are concerned. However, a discrepancy appears to exist between the way expressions of emotion are studied in the scientific literature, often using posed photographs, and actual expressions of emotion in reality, which are dynamic, spontaneous and potentially multimodal (combining visual expressions of face and body with emotional variations in speech). This is a potential problem for practical applications, since an emotion recognizer trained on posed expressions may perform poorly when tested on spontaneous expressions of emotion (e.g., Vogt and André, 2008). Therefore a better understanding of how posed expressions relate to non-posed, more spontaneous ones is called for.

In this chapter, we report on a series of experiments that compare posed and non-posed expressions of emotion, using minimal dynamic pairs collected using a language-based emotion induction procedure in which participants produce sentences of an increasingly strong emotional content. Some participants in these studies simply produce the sentences and the intended emotion is induced in them, while others are asked to produce the sentences while expressing an emotion opposite to the one expressed in the sentences. We study how the production of these sentences influences self-reported emotional states, and how the emotional expressions are perceived by others. In addition, since producing emotional sentences while expressing a different emotion might be easier for skilled actors, we explicitly compare facial expressions of inexperienced actors with those of experienced ones. We end with a brief discussion of other emotion induction techniques, and their usefulness for the study of non-verbal expressions of emotion.

6.1.1 Background

Modern work on facial expressions of emotion dates back to at least the seminal work of Darwin (1872) and, somewhat more recently, Ekman (1972). Many of the studies in this research tradition take the following form: pictures of facial expressions of basic emotions are presented to participants, who indicate which emotion is signaled. What these studies have repeatedly shown is that participants (or "judges") are capable of classifying these expressions well above chance, irrespective of age or culture (e.g., Elfenbein and Ambday, 2002; Russell et al., 2003; Schmidt and Cohn, 2002). The auditory expression of emotion has been less important historically, but here essentially the same results are obtained (e.g., Scherer, 2003), although agreement between judges is generally higher for the facial than for the vocal expressions (Walbott and Scherer, 1986). The combined vocal-facial expressions have so far received relatively little attention, with a few notable exceptions such as de Gelder and Vroomen (2000), Massaro and Egan (1996), and Scherer and Ellgring (2007).

In recent years it has been pointed out that the general methodology applied in these earlier studies has certain limitations. First of all, for practical reasons, most of the studies are based on static images, while in reality people normally make complex judgments based on "fleeting, dynamic signals encountered together with body posture and voice stimuli" (Adolphs 2002, p. 54). Moreover, and again for practical reasons, the actual emotional state of the person who displays the expression (the "sender") is usually not taken into account. Instead, senders are usually asked to act as if they are in a certain emotional state and to convey the corresponding emotions via their face. Often this is done in an iterative way, where portrayals that achieve the highest agreement between judges are understood as the correct signals. It has been argued that this is a method of "dubious ecological validity" (Russell et al., 2003, p. 333). In research on vocal expressions of emotion, using simulated (portrayed) expressions has also been the preferred way for collecting emotional voice data (Scherer, 2003, p. 232).

One important advantage of using actors is that it is generally easier to instruct an actor to display particular emotions than it is to induce them directly in participants, and ethical issues are no stumbling block (which is especially relevant for negative emotions). However, in general, it is fair to say that we still do not know in sufficient detail how posed and spontaneous expressions of emotion relate to each other, even though it is acknowledged that a better understanding of this relation is needed. Scherer (2003, p. 47), for example, writes that "obviously, one has to carefully investigate to what extent such acted material corresponds to naturally occurring emotional speech." He adds that there have been only few studies in which a systematic attempt has been made to compare portrayed and naturally occurring expressions of emotion. Of these studies a number were devoted to smiles.

It is well-known that people may display smiles for different reasons; they may produce spontaneous smiles as an expression of happiness – the so-called Duchenne smile –, or posed ones which may be employed for, for instance, social reasons – the non-Duchenne smile – (see e.g., Ekman et al., 1988; Frank et al., 1993; Fridlund, 1991; Hess and Kleck, 1990, 1994; Hess et al., 1992; Schmidt et al., 2003, 2006). Interestingly, it appears that spontaneous smiles differ from posed ones in that they generally have a smaller amplitude and a longer onset, presumably as a result of different Zygomatics Major muscle action (e.g., Cohn and Schmidt, 2004; Schmidt et al. 2003, 2006). On the perception side differences between the posed and spontaneous smiles have been found as well (e.g., Hess and Kleck, 1990, 1994). Apparently, children are less sensitive to certain markers for posed and spontaneous smiles than adults (e.g., del Giudice and Collie, 2007; Gosselin et al., 2002), and only gradually learn to distinguish between the two kinds of smiles. Fridlund (1991) looked explicitly at the effect of audience, both actual and implicit ones, on the production of joyful smiles, showing that they varied as a function of social context (estimated by facial electromyography or EMG). This result contributed to the development of Fridlund's (1994) behavioral ecology view of expressions (an alternative to the view that facial expressions are signs of emotional states). All these studies focus on visual smiles, but in a recent study, Drahota et al., (2008) show that there are also auditory differences between speech produced during Duchenne and non-Duchenne smiles. A number of studies have also compared posed and spontaneous expressions of other emotions besides happiness. These include Motley and Camden (1988) and Hess and Kleck (1994) who compared posed and spontaneous expressions of a range of positive and negative emotions. In general, it is worth emphasizing that much of these earlier works are based on static photographs and not on dynamic expressions. In addition, often data sets are compared that differ along other dimensions besides posed-spontaneous as well. As a result, even though there is suggestive evidence that posed and spontaneous expressions of the same emotion may differ, we still have an incomplete understanding of the exact relation between the two.

6.1.2 The Current Studies

Contra si gestus ac vultus ab oratione dissentiat, tristia dicamus hilares, adfirmemus aliqua renuentes, non auctoritas modo verbis sed etiam fides desit.[1]
Marcus Fabius Quintillianus

In this chapter, we describe a series of experiments systematically comparing posed and non-posed dynamic expressions of different emotions. The focus of this chapter is on facial expressions of emotion, but the methodology was chosen in such a way that vocal and audiovisual comparisons are possible as well, and we will return to this in the final Discussion.

As our starting point we use a novel adaptation of the Velten technique (Velten, 1968). The original Velten method is an example of a technique to experimentally induce emotions in participants, and according to Scherer (2003, p. 48) such methods have been "rather neglected in this research domain" (the vocal/visual expression of emotion). The basic idea of the Velten technique is that emotions can be induced in participants by letting them read a series of sentences that have a progressively stronger emotional content. It thus trades on the assumption that language and emotion are closely related (language can influence emotional states, and emotional states can influence language), an assumption that is currently gaining currency again (e.g., Beukenboom and Semin, 2006; Feldman et al., 2007; Stapel and Koomen, 2005). According to a meta-review of Westermann et al. (1996), the Velten technique was, at that time, "by far the most widely used induction method", although since the mid-nineties the film method has replaced the Velten technique as the induction method of choice for many researchers in the field. However, for our current purposes, the Velten technique seems especially useful since it directly involves speech production, and thus naturally allows for facial, vocal and audiovisual analyses.

The original Velten method was used to induce two specific emotions, to wit "elated" (joy) and "depressed" in Velten's terminology. In terms of the dimensional approach to emotions (e.g., Bachorowski, 1999) these two differ primarily along the valence dimension (positive and negative). One issue with induction methods such as these concerns so-called demand effects (e.g., Finegan and Seligman, 1995);

[1]English translation: "On the other hand, if gesture and the expression of the face are out of harmony with the speech, if we look cheerful when our words are sad, or shake our heads when making a positive assertion, our words will not only lack weight, but will fail to carry conviction" (Quintilianus, 1958, 11.3.67).

participants read sentences with a particular emotional content and might start displaying that particular emotion because they suspect this is what the experimenter "demands" of them. Naturally, this is a general risk for all experimental emotion induction procedures (except "unconscious" ones, e.g., Dimberg et al., 2000; Ruys and Stapel, 2008). In the context of the Velten method, these demand characteristics have been studied explicitly by instructing people "to behave as if a certain mood had been induced" (Buchwald et al., 1981), or by telling them that the induction would have an effect just opposite to that expected (Polivy and Doyle, 1980). Finegan and Seligman (1995) conclude that in the normal Velten conditions there is no evidence for demand characteristics. What is most relevant for our purposes, however, is that the way to study potential demand characteristics also offers a neat, minimal way to study the effects of posing on expressions of emotion. In particular, in addition to the two normal, non-posed Velten conditions, in which participants produce sentences that are congruent with the induced emotion, we add two posing conditions in which participants are explicitly instructed to utter the sentences in a way that is incongruent with their content.

In Experiment I, we use the Velten method in this way to elicit posed (incongruent) or non-posed (congruent) emotional expressions, in different experimental conditions. To find out how participants felt after doing this, they are asked to fill in an established questionnaire. If the Velten method works in our set-up, we would expect that participants in the congruent conditions indeed feel positive or negative afterwards (depending on the particular condition). The first question addressed in this chapter is what the self-reported emotional states of participants producing incongruent expressions will be. There is a popular belief that displaying certain emotions leads to feeling them ("Sit all day in a moping posture, sigh and reply to everything in a dismal voice, and your melancholy lingers," James, 1884), and various experimental studies have confirmed this effect (e.g., Stepper and Strack, 1993). On the other hand, there is also evidence that continuously displaying a smile, e.g., in occupations requiring constant cheerfulness (stewardesses, hamburger salespersons), does not lead to systematic positive feelings (Kotchemidova, 2005). Given these differing opinions in the literature, it will be interesting to see how participants in our incongruent conditions will feel afterwards.

The next question is whether differences in perception exist between the congruent and incongruent expressions. Given the reported differences between posed and spontaneous smiles, the question is whether such differences are perceptually relevant, and whether any perceptual differences also generalize beyond smiles to joyful expression in general and to other emotional expressions. Experiment II is conducted to find out whether gradual differences in perceived strength exist between congruent and incongruent expressions.

In Experiment I naive participants are asked to display emotions they do not feel, and which are incongruent with the content of the sentences they have to produce. It is conceivable that this requires good acting skills, and we therefore hypothesize that incongruent emotional expressions of professional actors are more realistic (and thus more similar to spontaneous expressions) than those of non-professional actors. To test this hypothesis, we redid the incongruent conditions from the first experiment in Experiment III, but this time with a group of experienced actors. In Experiment

IV, finally, the perception of the emotional expressions of these professional actors is compared with those of the non-professional actors (and with the non-acted ones).

6.2 Experiment I

In Experiment I we collect posed (incongruent) and non-posed (congruent) expressions of both a positive and a negative emotion, in addition to a neutral comparison condition, using a language-based induction method. After the induction, participants indicate how they feel at that moment, and the question is how participants in the incongruent condition will feel.

6.2.1 Method

Participants Participants were 50 students and colleagues from Tilburg University (31 females, 19 males), aged between 19 and 52, and with a mean age of 27 years. None of the participants was a professional or amateur actor, and none was involved with research on audiovisual speech or emotions. All participants gave written consent to use their data for research purposes, and none objected to being recorded. Participants were randomly assigned to one of the conditions.

Stimuli The sentences used in the various conditions were derived from the original set of sentences used by Velten, consisting of 180 sentences evenly distributed over three conditions (positive, negative and neutral). For our own experiments, positive and negative sentences were first literally translated in Dutch, after which they were revised to make sure they were easy to pronounce. Sentences from the original set that referred to "specifics" (e.g., college, parents, religion) were omitted. The neutral sentences ("There is a large rose-growing center near Tyler, Texas") were replaced with comparable sentences tailored towards the Dutch situation. In the end we selected 40 sentences for each condition. We made sure that the 40 sentences in the positive and negative condition showed the same progression as the original sets of 60 sentences, from neutral ("Today is neither better nor worse than any other day") to increasingly more emotional sentences ("God I feel great!" and "I want to go to sleep and never wake up." for the positive and negative sets, respectively), to allow for a gradual build up of the intended emotional state.

Procedure Participants took part one at a time. They were invited to a quiet room, where they were asked to take a seat in front of a desk on which a laptop computer was placed. The laptop was lifted 13 cm from the desk's surface so that the screen was more or less at eye level. Right above the screen a digital camera was positioned that recorded the face and upper body of the participants. Participants were told that the camera was only there to check afterwards whether the experimental procedure was properly followed.

Besides the three conditions described by Velten for the induction of congruent emotions (Positive Congruent, Neutral, Negative Congruent), two incongruent conditions were added. In one of these, participants were shown the negative sentences and were asked to utter these as if they were in a positive state (Positive Incongruent), in the other, positive sentences were shown and participants were instructed to utter these in a negative way (Negative Incongruent).

The instructions for the congruent and neutral conditions were a slightly abridged version of the original instructions from Velten. In the instruction phase of the congruent conditions, participants were told that the sentences would represent a "particular emotion" which was not further specified. They were asked to try and "experience" the contents of the sentences, as they were instructed to do in the original Velten method. In the incongruent conditions, participant were told that they would see sentences with a specific emotional content – "sentences radiating positivity and joy" or "sentences radiating somberness and depression", depending on the condition. They were then instructed to ignore this emotional content, and express the sentences as if they were in respectively a depressed or a joyful state. Other than that, the instructions were exactly the same as those for the congruent conditions. In the instructions for the neutral condition participants were merely asked to read each sentence twice, once silently and once out loud. It is important to stress that in none of the instructions for the individual conditions any reference was made to facial or vocal expressions of emotion.

Participants were told that the goal of the experiment was to study the effect of emotion on memory recall (earlier work has revealed that the effectiveness of induction procedures increases when the induction serves a clear purpose, e.g., Westermann et al., 1996). The instructions were displayed on the computer screen, and participants were instructed to first silently read the texts, after which they had to read them aloud. This enabled them to practice the experimental procedure. The introduction phase was self-paced.

If the instructions were clear, the experimenter left the room and the actual experiment started. During this phase, the sentences were displayed on a computer screen for 20 s, and participants were instructed to read each sentence twice (once silently, then out loud). This phase lasted exactly 800 s (40 sentences × 20 s), i.e., a little over 13 min. During the induction, participants were alone in the room, to avoid presence effects (Wagner and Lee, 1999).

Immediately following this phase, participants had to fill in a short self-report emotion questionnaire ("At this moment, I feel ...") derived from Mackie and Worth (1989) and adapted to Dutch in Krahmer, van Dorst and Ummelen (2004), consisting of six 7-point bipolar semantic differential scales, using the following adjective pairs (English translations of Dutch originals: happy/sad, pleasant/unpleasant, satisfied/unsatisfied, content/discontent, cheerful/sullen and in high spirits/low-spirited). The order of the adjectives was randomized; for processing, negative adjectives were mapped to 1 and positive ones to 7. The internal consistency of the emotion questionnaire was measured using Cronbach's α and was very good (α = 0.92).

After filling in the questionnaire participants performed a dummy recall test, as this was supposed to be the purpose of the mood induction. The results of the recall test were not analysed. Finally, participants were debriefed and told about the real purpose of the experiment. They were given a small gift as a token of appreciation.

Design Experiment I had a between-participants design, with Condition as independent variable (levels: Positive Congruent, Positive Incongruent, Neutral, Negative Incongruent, Negative Congruent) and with the self-reported emotion scores as the dependent variable.

6.2.2 Results and Discussion

Figure 6.1 contains a number of representative stills, while Table 6.1 reveals the induced emotional state on a 7-point scale (1 = very negative, 7 = very positive) as a function of condition. The congruent conditions clearly have the strongest

Fig. 6.1 Representative stills of Congruent (*top*) and Incongruent emotional (*bottom*) expressions, with on the *left hand side* the positive and on the *right hand side* the negative versions

Table 6.1 Mean self-reports of induced emotional state on a 7-point scale (1 = very negative, 7 = very positive) as a function of condition (standard deviations between brackets), with 95% confidence intervals

Condition	Induced emotion (SD)	95% CI
Positive congruent	5.65 (0.63)	(5.04, 6.23)
Positive incongruent	4.77 (1.23)	(4.16, 5.37)
Neutral	4.95 (0.87)	(4.34, 5.56)
Negative incongruent	4.92 (0.63)	(4.31, 5.52)
Negative congruent	3.85 (1.20)	(3.24, 4.65)

impact: participants in the Positive Congruent condition feel the most positive ($M = 5.65, SD = 0.63$) and participants in the Negative Congruent condition feel most negative afterwards ($M = 3.85, SD = 1.19$). The incongruent conditions result in essentially the same emotion as reading the Neutral condition (i.e., a neutral one). An analysis of variance (ANOVA) confirmed that condition had a significant effect on the self-reported emotional state of the participants, $F(4,45) = 4.65, p < 0.005, \eta^2 = 0.288$. A Tukey HSD post hoc analysis revealed that the scores for the Positive Congruent and Negative Congruent conditions differed significantly, but none of the other pairwise comparisons did.

The first experiment revealed that the language-based emotion induction method in this set-up worked as intended; in the congruent conditions the intended emotional states were indeed induced, which shows that translating the set of Velten sentences and reducing it from 60 to 40 sentences per condition did not have a negative effect on the usefulness of the method. The first central question in this chapter was how participants in the incongruent conditions would feel afterwards, and it is interesting to observe that participants in the incongruent conditions felt on average the same as those in the neutral condition, indicating that the incongruent expressions were not felt. Next, we turn to the perception of the congruent and incongruent expressions.

6.3 Experiment II

In this experiment, we investigate whether perceptual differences exist between the congruent and incongruent dynamic expressions collected for Experiment I exist.

6.3.1 Method

Participants Forty people participated (all different from those of the first experiment), 20 females and 20 males, with an average age of 36.

Stimuli The stimuli used in this experiment consist of the last utterance for each of the 50 participants from Experiment I. These utterances, uttered just before filling in the questionnaire, arguably capture the speakers at the height of the induced emotion. The utterances were cut from just before the participant starts speaking, to just after the sentence was finished. Stimuli were offered in a vision-only format (without sound), to prevent participants from relying on lexical information to make their choice.

Procedure Participants took part one at a time. They were invited into a quiet room, and asked to take place in front of a computer. Participants were told that they would see 50 speakers in different emotional states, and that their task was to rate the perceived state on a 7 point valence scale ranging from 1 (= very negative) to 7 (= very positive). Participants were not informed about the fact that some of the speakers were expressing incongruent emotions. The stimuli were offered in one of two random orders, to compensate for potential learning effect. They were preceded by a number displayed on the screen indicating which stimulus would come up next, and followed by a 3 s interval during which participants could fill in their score on an answer form. Stimuli were shown only once. The experiment was preceded by a short training session consisting of three speakers (for which a different sentence was used) to make participants acquainted with the stimuli and task. If all was clear, the actual experiment started, after which there was no further interaction between participant and experimenter. The entire experiment lasted approximately 10 min.

Design Experiment II had a within-participants design, with Condition as independent variable (levels: Positive Congruent, Positive Incongruent, Neutral, Negative Incongruent, Negative Congruent) and with the perceived emotion scores as the dependent variable.

6.3.2 Results and Discussion

Table 6.2 summarizes the results. A repeated measures analysis of variance (ANOVA) shows that condition has a significant effect on perceived valence, $F(4, 156) = 472.79, p < 0.001, \eta^2 = 0.92$, after a Greenhouse-Geisser correction.[2] Post hoc analyses using the Bonferroni method show that all conditions lead to a significantly different perceived emotional state (at $p < 0.001$), with the sole exception of the difference between Positive Congruent and Positive Incongruent. It is interesting to observe that the Incongruent conditions are perceived more strongly than the real ones; speakers in the Positive Incongruent condition are perceived as the most positive ($M = 4.86, SD = 0.40$) although the difference with the speakers in the Positive Congruent is small ($M = 4.81, SD = 0.36$), and speakers in the Negative Incongruent condition are perceived as the most negative ($M = 2.54, SD = 0.50$).

[2] Here and elsewhere, we report on the normal degrees of freedom and error values after such a correction, for the sake of readability.

Table 6.2 Mean perceived emotional state scores on a 7-point scale (1 = very negative, 7 = very positive) as a function of condition (standard errors between brackets), with 95% confidence intervals

Condition	Perceived emotion (SE)	95% CI
Positive congruent	4.81 (0.06)	(4.70, 4.93)
Positive incongruent	4.86 (0.06)	(4.73, 4.99)
Neutral	3.52 (0.07)	(3.38, 3.66)
Negative incongruent	2.54 (0.08)	(2.38, 2.70)
Negative congruent	3.06 (0.07)	(2.93, 3.19)

This perception experiment revealed that seeing speakers producing incongruent emotional expressions leads to more extreme perceived valence scores than seeing speakers produce congruent expressions, where the difference between congruent and incongruent expressions is particularly strong for the negative conditions

6.4 Experiment III

The evidence gathered so far suggests that incongruent expressions of emotion (in which participants display an emotion which they do not feel) are perceived more strongly than congruent expressions of emotion (in which participants do not pose). However, the participants in the incongruent conditions of Experiment I were not trained actors, so it might be that they display the emotional expressions in a stronger way than congruent emotions, simply because their acting capabilities are not sufficiently well-developed. In Experiments III and IV we investigate to what extent acting experience impacts the production and perception of incongruent expressions of emotion. For this we collected additional data from 20 experienced actors. The hypothesis is that incongruent expressions from professional actors will be more like congruent expressions than the incongruent expressions from non-professional actors. In Experiment III we describe this data collection and compare the self-reported emotional state scores with those of the participants in the positive and negative conditions collected for Experiment I.

6.4.1 Method

Participants Twenty professional actors participated (besides the speakers already collected for Experiment I), either experienced actors from various theater companies or students in the final year of the drama academy. All had between 3 and 25 years of professional experience ($M = 11.2$ years, $SD = 6.5$ years). Ten actors were female, ten male. All participants gave written consent to use their data for research purposes, and none objected to being recorded. Participants were randomly assigned to one of the two incongruent conditions.

Stimuli The stimuli sentences were identical to those used in Experiment I.

Procedure The procedure in the respective conditions was identical to that of Experiment I. Crucially, the professional actors received exactly the same instructions as the non-professional acting participants in Experiment I, and they did not know that their acting skills were of interest to the experiment. They were only informed about this after the experiment was finished, during the debriefing. Again, the internal consistency of the emotion questionnaire was measured using Cronbach's α and was very good ($\alpha = 0.93$).

Design Experiment III had a between participants design with Acting (3 levels: No-acting, Inexperienced-acting and Experienced-acting) and Valence (2 levels: Positive, Negative) as the independent variables, and self-reported emotional state as the dependent variable. The No-acting condition consists of the Congruent conditions and the Inexperienced acting consists of the Incongruent conditions from Experiment I.

6.4.2 Results and Discussion

Figure 6.2 shows a number of representative stills of experienced actors in the Positive and Negative (Incongruent) conditions. Figure 6.3 depicts the self-reported emotional state scores from the experienced actors in the Positive and Negative conditions, and compares them to the scores from participants who did not act (those in the congruent conditions of Experiment I) and from inexperienced acting participants (the incongruent conditions from Experiment I).

An Analysis of Variance (ANOVA) revealed a main effect of Valence, $F(1,54) = 12.543, p < 0.001, \eta^2 = 0.188$. Overall participants in Positive conditions felt more positive afterwards than participants in Negative conditions. No main effect of Acting was found ($F < 1$). However, a significant interaction between Valence and Acting was found, $F(2,54) = 5.201, p < 0.01, \eta^2 = 0.162$. This interaction is readily explained by inspection of Fig. 6.3: in the No-acting (Congruent) condition the self-reported emotion scores between participants in the Positive and Negative Congruent conditions are most different, while the differences between Inexperienced participants are negligible. Interestingly, the scores for the experienced actors in the Positive and Negative (Incongruent) conditions are almost exactly in between these two extremes: actors in the Positive Incongruent condition, indicate that they feel somewhat more positive at the end of the experiment ($M = 5.32, SD = 1.06$) than actors in the Negative Incongruent condition ($M = 4.35, SD = 0.77$).

Even though Acting skills did not have a significant main effect in Experiment III, there was a significant interaction between Acting and Valence; while our inexperienced actors indicated that they felt neutral afterwards (Experiment I), we did find a small, significant effect on how the experienced actors indicated they felt afterwards, with actors in the Positive Incongruent condition reporting higher scores than those

Fig. 6.2 Representative stills of incongruent emotional expressions from male (*top*) and female (*bottom*) experienced actors, with on the *left hand side* the positive and on the *right hand side* the negative versions

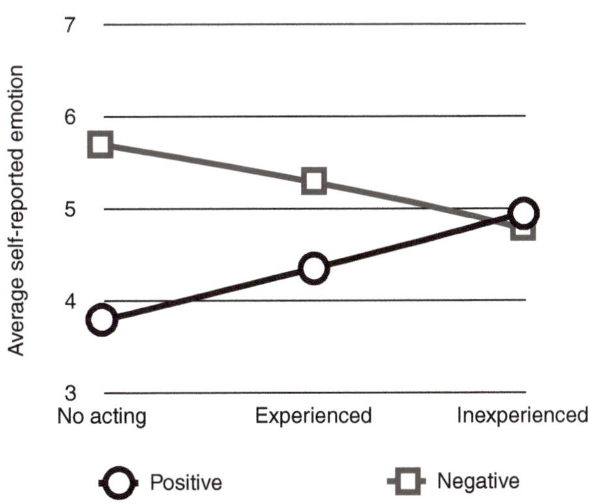

Fig. 6.3 Self-reported emotional state scores for Non-actors (congruent conditions, Experiment I), Inexperienced actors (incongruent conditions, Experiment I) and Experienced actors (incongruent conditions, Experiment III), as a function of valence (*Positive*, *Negative*)

in the Negative Incongruent condition. The crucial question is of course how the utterances from these experienced actors were perceived, especially in comparison with the other speakers collected for Experiment I, where the incongruent expressions from professional actors (but not from non-professional ones) are expected to be similar to congruent expressions. This is addressed in Experiment IV.

6.5 Experiment IV

Experiment IV is a replication of Experiment II, with the 20 speakers collected for Experiment III (the professional actors) added.

6.5.1 Method

Participants Forty people participated as judges, 20 male and 20 female ones, with an average age of 36.2 years. None had participated in Experiments I–III.

Stimuli For each of the speakers who participated in Experiment I or in Experiment III the final utterance was selected and processed as described above, giving rise to 70 stimuli. Stimuli were again presented in a vision-only format, to prevent participants from relying on lexical information, and were shown to participants in one of two random orders to compensate for potential learning effects.

Procedure The procedure was identical to that of Experiment II, but naturally, the addition of 20 stimuli lengthened the duration of the experiment with a few minutes.

Design Experiment IV had a within participants design with Condition as independent variable (levels: Positive Congruent, Positive Incongruent (inexperienced actors), Positive Incongruent (experienced actors), Neutral, Negative Incongruent (inexperienced actors), Negative Incongruent (experienced actors), Negative Congruent) and with the perceived emotion scores as the dependent variable.

6.5.2 Results and Discussion

Table 6.3 summarizes the results. A repeated measures analysis of variance (ANOVA) revealed a significant effect of Condition on perceived emotional state, $F(6, 234) = 360.465, p < 0.001, \eta^2 = 0.902$. Pairwise comparisons (after a Bonferroni correction) revealed that all conditions were significantly differently perceived ($p < 0.001$), except the comparison between Positive Congruent and Positive Incongruent (by Inexperienced Actors). The emerging picture is surprisingly consistent. Speakers in the Positive conditions are perceived as more positive

Table 6.3 Mean perceived emotional state scores on a 7-point scale (1 = very negative, 7 = very positive) as a function of condition (standard errors between brackets), with 95% confidence intervals. Two variants of the incongruent conditions are included, one with Inexperienced Actors and one with Experienced actors

Condition	Perceived emotion (SE)	95% CI
Positive congruent	4.69 (0.07)	(4.56, 4.82)
Positive incongruent inexperienced actors	4.70 (0.07)	(4.56, 4.84)
Positive incongruent experienced actors	5.71 (0.09)	(5.53, 5.89)
Neutral	3.56 (0.07)	(3.41, 3.70)
Negative incongruent inexperienced actors	2.89 (0.10)	(2.69, 3.09)
Negative incongruent experienced actors	2.40 (0.09)	(2.21, 2.59)
Negative congruent	3.29 (0.07)	(3.15, 3.43)

and those in the Negative conditions as more negative, with neutral precisely in between. Interestingly, in all cases the incongruent conditions are perceived more strongly than the Congruent ones, albeit that the difference between Positive Congruent ($M = 4.69, SD = 0.40$) and Positive Incongruent (inexperienced actors) ($M = 4.70, SD = 0.43$) is, as in Experiment II, insignificant. And, most interestingly, the stimuli of the Experienced Actors receive the most extreme scores ($M = 5.71, SD = 0.56$ for Positive Incongruent and $M = 2.40, SD = 0.60$ for Negative Incongruent), where the difference with the scores for the Inexperienced Actors is quite substantial, especially for the Positive conditions.

Experiment IV confirmed the findings of Experiment II: seeing speakers producing incongruent emotional speech leads to more extreme perceived emotion scores than seeing speakers produce congruent emotional speech. In addition: Experiment IV revealed a clear and consistent effect of acting experience, but not in the way it was expected: when the incongruent expressions are produced by professional actors, their utterances are perceived significantly stronger than those produced by their inexperienced counterparts. Finally, it is noteworthy that, even though Experiment IV replicates the findings of Experiment II, the scores for the 5 conditions present in both Experiments are pushed a little bit more to the centre of the scale, presumably as a side effect of the 20 additional speakers which are perceived as more "extreme" than the other 50.

6.6 General Conclusion and Discussion

We have described a series of experiments, comparing congruent (spontaneous) expressions of emotions with incongruent (posed) ones.

6.6.1 Discussion of the Results

Experiment I revealed that congruent expressions have a stronger impact on the self-reported emotion scores; participants that produce incongruent sentences feel (close to) neutral afterwards, while participants that produce positive or negative congruent sentences indeed feel more positive or negative. This also means that the current adaptation of the Velten technique (reduced from 60 to 40 sentences, and translated to Dutch) turned out to be an effective induction procedure. Since the technique is purely language based, it further emphasizes the observation that language can influence emotional state. The first question raised above was whether posed and non-posed expressions have a different impact on how participants feel afterwards, and this was indeed the case. Little evidence was found for the claim that displaying certain emotions leads to feeling them, which is in line with what is known from the way professional actors (both in Europe and in the US) play emotions (Konijn, 2000).

Experiment II looked at perceptual differences between the various dynamic congruent and incongruent expressions of the respective emotions, where it was found that the incongruent expressions are perceived as stronger than the congruent counterparts. This is consistent with the suggested differences between different kinds of smiles (e.g., Cohn and Schmidt, 2004), where the current experiments show that such differences are perceptually relevant, and generalize to dynamic facial expressions of happiness, and also to the negative emotion under study here. In general, this suggests that emotions that are felt are not always fully displayed (as was also suggested by Reisenzein et al., 2006 for the case of surprise) while emotions that are not felt are more fully, and more stereotypically displayed. This general pattern (incongruent expressions of emotion have a lesser impact on how the speaker feels, but are perceived more strongly) was fully confirmed in a replication with speakers from a different, south-Asian culture (see Shahid et al., 2008a).

Finally, we hypothesized that expressions of professional actors would be more realistic (more like congruent expressions) than those of non-professional actors. After all, one possible explanation of the differences found between congruent and incongruent expressions in the first three experiments is that the participants in the incongruent conditions simply did not have sufficient acting skills. Displaying an emotional expression that is contradicted by the semantic content of the sentences to be produced is perhaps not a natural task and may require certain acting skills. To rule out this explanation we ran two replication experiments (Experiments III and IV) with experienced actors as participants. These experiments fully confirmed our initial findings, and contrary to our expectations, it turned out that the facial expressions of the experienced actors were perceived as even more extreme than those of the participants without an education in and professional experience with acting. Naturally, it can be claimed that if the actors would be trained using, say, the Stanislavski method or if they had a background in method acting they might display more subtle expressions (e.g., Scherer, 2003; Marsella et al., 2006), or alternatively it can be pointed out that expressions that are elicited using extensive scenarios (Enos and Hirschberg, 2006) would be more realistic (and note that such procedures are currently becoming more popular in emotion elicitation studies involving actors). But one might expect that expressions from experienced actors would at least go some way in the more realistic direction, which is not what we found. In addition, it is worth noting that in most of the earlier studies using posed expressions, the authors are typically vague about how the posed expressions were elicited, and it seems a safe bet that the elicitation procedure did not involve method acting, nor extensive scripting.

6.6.2 On Posed Expressions

The evidence gathered in these experiments suggests that incongruent (posed) expressions of emotion are more intense and more proto-typical than congruent (non-posed) ones; hence the more extreme scores in Experiments II and IV. If we concede that incongruent (posed) expressions indeed are more intense and prototypical (like "caricatures", Feldman Barret et al., 2007, p. 328) than congruent (non-posed) expressions, one logical follow-up question is what the status of these expressions is. Arguably, the intense and prototypical expressions capture something important and easily recognizable about emotional expressions. However, if the goal is to (automatically) recognize or understand real human expressions of emotion (e.g., Pantic and Patras, 2006 for facial expressions and Vogt and André, 2005 for speech) knowledge based on posed expressions may not be very useful, since we have found that the posed expressions do differ from their non-posed counterparts. Probably, this difference is caused by at least two related observations. First of all, in the incongruent conditions, there might be some amount of dislocation between feeling and displaying; participants indicate that they "feel" more joyful or depressed, but they do not display this on the face. This is similar to what Reisenzein et al. (2006) found for "surprise". They also present evidence for a dislocation between feeling and displaying surprise; participants indicated that they were surprised (self reports), but surprise expressions were rare (and were usually "incomplete", not containing all ingredients of a typical surprise expression). And this brings us to the second point: it appeared that the incongruent emotional expressions more often contain the full, stereotypical expressions. To confirm this observation, we performed an annotation of the visual cues in the stimuli, which indeed revealed that the stereotypical emotional expressions (pronounced smile, raised brows, etc.) are mostly found in the incongruent cases, while the congruent ones are mostly incomplete (e.g., only a vague smile), which is in line with the claims from, for instance, Galati et al. (1997) and Horstman (2002) that people do not frequently display entire stereotypical expressions spontaneously. The interested reader is referred to Barkhuysen et al. (2010) for a more complete description of the facial features associated with the different congruent and incongruent conditions.

6.6.3 Outlook

Arguably, the studies that we have described in this chapter have two limitations. One limitation concerns the production experiments, where participants were either asked to read emotional sentences while displaying an opposite emotion or were asked to simply experience the emotion contained in the sentences. While we assume that congruent sentences are produced with spontaneous expressions of emotions, and incongruent ones with posed expressions, we cannot be absolutely sure that this is indeed the case. For example, it is conceivable that some participants in the congruent conditions are purposefully making emotional faces in order to try and get into their role of reader and "experience" the emotion. Thus, some may have displayed congruent, but nevertheless posed expressions. Second, although our assumption was that in the absence of instructions (i.e., not being told to show a different emotion) all emotions displayed will be congruent ones, it is possible that a small minority of the participants might have had a spontaneous but opposite response while reading the emotional sentences (e.g., if they felt embarrassed in reading a statement about "how great they were feeling"). In this case, the responses would be incongruent with the content of the sentences, but spontaneous in its occurrence. Even though the results of the perception studies lend no support for these hypothetical situations, they cannot be ruled out with certainty.

Another limitation concerns the perception experiments, namely that they are based on the facial expressions only. Presenting the recordings to participants (judges) in an audiovisual format is complicated since the lexical material of the sentences is a give-away clue for the emotional state of the speaker. It is interesting to observe that the incongruent expressions appear to be "ironic" (imagine someone saying "God I feel great!" in a depressed tone of voice). In fact, this is particularly true for the negative incongruent condition (and to a lesser extent for the positive incongruent one). This perception of irony is not surprising, since it is precisely the mismatch between form and content that triggers the ironic interpretation (compare also the Quintillianus quote at the beginning of this chapter). However, since we are also interested in the perception of the recordings in the vocal and audiovisual context an additional series of perception experiments was conducted with foreign language speakers. In particular, Czech participants were asked to rate the perceived emotional state of the 50 speakers collected in Experiment I, in one of three conditions: vision-only (a replication of Experiment II), audio-only and audio-visual. These experiments, discussed in more detail in Barkhuysen et al. (2010), revealed similar results as described in this chapter (incongruent expressions perceived stronger than congruent ones) for all three modalities, although the differences in the visual condition were more pronounced than those in the auditory one.

In this chapter we have seen that the Velten method not only offers interesting possibilities to study posed and spontaneous expressions of emotion, but is first and foremost a useful method to induce emotions in people. It is interesting to compare this method with two other popular induction methods discussed in the aforementioned meta-study of Westermann et al. (1996): the film method and the feedback method. The film method is based on the notion that film fragments with a strong emotional content induce similar feelings in those watching the fragments. In Swerts and Krahmer (2008) the film method was used to study differences in the production and perception of audiovisual emotional expressions by male and female speakers. After watching a 7 min film fragment with either a positive or a negative valence, participants were interviewed about the fragment they saw, and the video recordings were used in further perception studies. The feedback method rests on the assumption that receiving positive or negative social feedback induces positive or negative feelings in participants. This method was used in Krahmer et al. (2008) to study non-verbal cues of social emotions in direct interactions. Participants engaged in a conversation in which they, for a limited period of time. were either included or excluded from an ongoing conversation. The recordings made in this way were subsequently used for a series of perception experiments. In these two studies the induced emotions were measured using the same self-report questionnaire as used for Experiments I and III in this chapter, which makes it possible to directly compare the effects of the three induction methods. Such a comparison reveals that the film method induces stronger and the feedback method induces weaker emotions in participants than our application of the Velten method did, even though in all studies the positive and negative conditions differed significantly from one another.

A method of a more recent date, and hence not discussed by Westermann et al. (1996), is using computer games as a way of inducing emotions. For example, Shahid et al. (2008b) used a simple (and "fake") card guessing game, which children would win and lose at predictable places to induce positive and negative emotions, while Shahid et al. (2009) experimented with a game based on a digital, interactive laughing mirror as a tool to induce joy in players. Both methods lend themselves very well for collecting audiovisual expressions of emotion, and have been used, among other things, to study the effect of physical co-presence on displaying emotions.

Although each of these methods has its own specific advantages – the Velten method seems especially useful for collecting speech samples with an emotional content, the film method is best suited for the induction of specific emotions (see e.g., Rottenberg et al., 2007), the feedback method appears to be particularly good for studying emotional expressions in face-to-face interactions, and the games method is very natural and hence suitable for emotion induction in child participants – all these methods proved to be very useful for the collection of dynamic, spontaneous and multimodal expressions of emotion.

Acknowledgements We are grateful to Emmett Velten for supplying us with the sentences used in his original experiment, and for some useful discussion on acting and demand effects. We would like to thank Lennard van de Laar, Marieke Rikken, Suleman Shahid, Pashiera Barkhuysen and, especially, Janneke Wilting for their help with and discussions about the experiments. Thanks are due to Harold Miesen and Carel van Wijk for statistical advice, and to Diederik Stapel, Kirsten Ruys and Maarten Brouwers for useful comments on a previous version of this chapter.

References

Adolphs, R.: Neural systems for recognizing emotion. Curr. Opin. Neurobiol. **12**, 169–177 (2002)

Bachorowski, J.: Vocal expression and perception of emotion. Curr. Dir. Psychol. Sci. **8**(2), 53–57 (1999)

Barkhuysen, P., Krahmer, E., Swerts, M.: Cross-modal and incremental perception of audiovisual cues to emotional speech. Lang. Speech, **53**, 3–30 (2010, in press)

Beukenboom, C., Semin, G.: How mood turns on language. J. Exp. Soc. Psychol. **42**, 553–566 (2006)

Buchwald, A., Strack, J., Coyne, J.: Demand characteristics and the Velten mood induction procedure. J. Consult. Clin. Psychol. **49**, 478–479 (1981)

Cohn, J., Schmidt, K.: The timing of facial motion in posed and spontaneous smiles. Int. J. Wavelets Multiresolut. Inf. Proc. **2**, 1–12 (2004)

Darwin, C.: The Expression of the Emotions in Man and Animals, 3rd edn. Oxford University Press, New York, NY [1998] (1872)

Del Giudice, M. and Colle, L. (2007). Differences between children and adults in the recognition of enjoyment smiles. Developmental Psychology, **43**, 796–803

Dimberg, U., Thunberg, M., Elmehed, K.: Unconscious facial reactions to emotional facial expressions. Psychol. Sci. **11**, 86–89 (2000)

Drahota, A., Costalla, A., Reddya, V.: The vocal communication of different kinds of smile. Speech Commun. **50**, 278–287 (2008)

Ekman, P.: Emotion in the Human Face. Pergamon Press, New York, NY (1972)

Ekman, P., Friesen, W., O'Sullivan, M.: Smiles when lying. J. Personal. Soc. Psychol. **54**, 414–420 (1988)

Elfenbein, H., Ambday, N.: On the universality and cultural specificity of emotion recognition: a meta-analysis. Psychol. Bull. **128**, 203–235 (2002)

Enos, F., Hirschberg, J.: A framework for eliciting emotional speech: capitalizing on the actor's process. In: L. Devillers, J.-C. Martin, R. Cowie, E. Douglas-Cowie and A. Batliner (Eds.) LREC 2006 Workshop on Corpora for Research on Emotion and Affect. Genova, Italy (2006)

Feldman Barett, L., Lindquist, K., Gendron, M.: Language as context for the perception of emotion. Trends Cogn. Sci. **11**(8), 327–332 (2007)

Finegan, J., Seligman, C.: In defense of the Velten mood induction procedure. Can. J. Behav. Sci. **27**, 405–418 (1995)

Frank, M., Ekman, P., Friesen, W.: Behavioral markers and recognizability of the smile of enjoyment. J. Personal. Soc. Psychol. **64**, 83–93 (1993)

Fridlund, A.: The sociality of solitary smiles: effects of an implicit audience. J. Personal. Soc. Psychol. **60**, 229–240 (1991)

Fridlund, A.: Human Facial Expression. An Evolutionary View. Academic, San Diego (1994)

Galati, D., Scherer, K., Ricci-Bitti, P.: Voluntary facial expression of emotion: comparing congenitally blind with normal sighted encoders. J. Personal. Soc. Psychol. **73**, 1363–1379 (1997)

de Gelder, B., Vroomen, J.: The perception of emotions by ear and by eye. Cogn. Emot. **14**, 289–311 (2000)

Gosselin, P., Perron, M., Legault, M., Campanella, P.: Children's and adults' knowledge of the facial features distinguishing genuine from simulated smiles. J. Nonverbal Behav. **26**, 83–108 (2002)

Hess, U., Kleck, R.: Differentiating emotion elicited and deliberate emotional facial expressions. Eur. J. Soc. Psychol. **24**, 367–381 (1990)

Hess, U., Kleck, R.: The cues decoders use in attempting to differentiate emotion-elicited and posed facial expressions. Eur. J. Soc. Psychol. **20**, 369–385 (1994)

Hess, U., Kappas, A., McHugo, G.J., Lanzetta, J., Kleck, R.: The facilitative effect of facial expression on the self-generation of emotion. Int. J. Psychophysiol. **12**, 251–265 (1992)

Horstmann, G.: Facial expressions of emotion: does the prototype represent central tendency, frequency of instantiation, or an ideal? Emotion **2**, 297–305 (2002)

James, W.: What is an emotion? Mind **9**, 190–204 (1884)

Konijn, E.: Acting Emotions: Shaping Emotions on Stage. Amsterdam University Press, Amsterdam (2000)

Kotchemidova, C.: From good cheer to "drive-by smiling": a social history of cheerfulness. J. Soc. Hist. **39**(1), 5–37 (2005)

Krahmer, E., van Dorst, J., Ummelen, N.: Mood, persuasion and information presentation. Inf. Des. J. **12**, 40–52 (2004)

Krahmer, E., Schaafsma, J., Swerts, M., Vingerhoets, A. Nonverbal responses to social inclusion and exclusion. In: Proceedings of the International Conference on Spoken Language Processing (Interspeech 2008), ISCA, Brisbane, pp. 809–812 (2008)

Mackie, D., Worth, L.: Processing deficits and the mediation of positive affect in persuasion. J. Personal. Soc. Psychol. **57**, 27–40 (1989)

Marsella, S., Carnicke, S., Gratch, J., Okhmatovskaia, A., Rizzo, A.: An exploration of Delsarte's structural acting system. In: Proceedings of the 6th International Conference on Intelligent Virtual Agents (IVA), Marina del Rey, CA (2006)

Massaro, D., Egan, P.: Perceiving affect from the voice and the face. Psychon. Bull. Rev. **3**, 215–221 (1996)

Motley, M., Camden, C.: Facial expression of emotion: a comparison of posed expressions versus spontaneous expressions in an interpersonal communication setting. West. J. Speech Commun. **52**, 1–22 (1988)

Pantie, M., Patras, I.: Dynamics of facial expression: recognition of facial actions and their temporal segments from face profile image sequences. IEEE Trans. Syst. Man Cybern. B **36**, 433–449 (2006)

Picard, R., Klein, J.: Computers that recognise and respond to user emotion: theoretical and practical implications. Interact. Comput. **14**(2), 141–169 (2002)

Picard, R., Vyzas, E., Healy, J.: Toward machine emotional intelligence: analysis of affective physiological state. IEEE Trans. Pattern Anal. Mach. Intell. **23**(10), 1175–1191 (2001)

Polivy, J., Doyle, C.: Laboratory induction of mood states through the reading of self-referent mood statements: affective changes or demand characteristics. J. Abnorm. Psychol. **89**, 286–290 (1980)

Quintilianus, M.F.: Instituţia Oratoria (trans. in English by Butler, H.E.). Harvard University Press, Cambridge (1958)

Reisenzein, R., Bördgen, S., Holtbernd, T., Matz, D.: Evidence for strong dissociation between emotion and facial displays: the case of surprise. J. Personal. Soc. Psychol. **91**, 295–315 (2006)

Rottenberg, J., Ray, R., Gross, J.: Emotion elicitation using films. In: Coan, J.A., Allen, J.J.B. (eds.) The Handbook of Emotion Elicitation and Assessment. Oxford University Press, London, 9–28. (2007)

Russell, J., Bachorowski, J., Fernandez-Dols, J.: Facial and vocal expressions of emotion. Ann. Rev. Psychol. **54**, 329–349 (2003)

Ruys, K., Stapel, D.: The secret life of emotions. Psychol. Sci. **19**, 385–391 (2008)

Scherer, K.: Vocal communication of emotion: a review of research paradigms. Speech Commun. **40**, 227–256 (2003)

Scherer, K., Ellgring, H.: Multimodal expression of emotion: affect programs or componential appraisal patterns? Emotion **7**, 158–171 (2007)

Schmidt, K., Cohn, J.: Human facial expressions as adaptations: evolutionary questions in facial expressions. Yearb. Phys. Anthropol. **44**, 3–24 (2002)

Schmidt, K., Ambadar, Z., Cohn, J., Reed, L.: Movement differences between deliberate and spontaneous facial expressions: zygomaticus major action in smiling. J. Nonverbal Behav. **3**, 37–52 (2006)

Schmidt, K., Cohn, J., Tian, Y.: Signal characteristics of spontaneous facial expressions: automatic movement in solitary and social smiles. Biol. Psychol. **65**, 49–66 (2003)

Shahid, S., Krahmer, E., Swerts, M.: Emotional speech: comparing South-Asian and Caucasian speakers and observers. In: Proceedings of the Fourth Speech Prosody Conference, Campinas, Brazil (2008a)

Shahid, S., Krahmer, E., Swerts, M.: Alone or together: exploring the effect of physical co-presence on the emotional expressions of game playing children across cultures. In: Markopoulus, P. et al. (eds.) Fun and Games. Lecture Notes in Computer Science LNCS, vol. 5294, pp. 94–105. Springer, Berlin/Heidelberg (2008b)

Shahid, S., Krahmer, E., Swerts, M., Melder, W.A., Neerinex, M.A. Exploring social and temporal dimensions of emotion induction using an adaptive affective mirror. In: Proceedings CHI 2009: Extended Abstracts, Boston, MA, pp. 3727–3732 (2009)

Stapel, D., Koomen, W.: When less is more. The consequences of affective primacy for subliminal priming effects. Personal. Soc. Psychol. Bull. **31**, 1286–1295 (2005)
Stepper, S., Strack, F.: Proprioceptive determinants of emotional and non-emotional feelings. J. Personal. Soc. Psychol. **64**, 210–22 (1993)
Swerts, M., Krahmer, E.: Gender-related differences in the production and perception of emotion. In: Proceedings of the International Conference on Spoken Language Processing (Interspeech 2008), ISCA, Brisbane, pp. 334–337 (2008)
Velten, E.: A laboratory task for induction of mood states. Behav. Res. Therapy **6**, 473–482 (1968)
Vogt, T., André, E.: Comparing feature sets for acted and spontaneous speech in view of automatic emotion recognition. In: Proceedings of IEEE International Conference on Multimedia & Expo (ICME 2005), Trento, Italy (2005)
Walbott, H., Scherer, L.: Cues and channels in emotion recognition. J. Personal. Soc. Psychol. **51**, 690–699 (1986)
Wagner, H., Lee, V.: Facial behavior alone and in the presence of others. In: Philippott, P., Feldman, R., Coats, E. (eds.) The social context of nonverbal behavior, pp. 262–286. Cambridge University Press, New York, NY (1999)
Westermann, R., Spies, K., Stahl, G., Hesse, F.: Relative effectiveness and validity of mood induction procedures. Eur. J. Soc. Psychol. **26**, 557–580 (1996)

Chapter 7
Accessing the Parallel Universe of Connotative Meaning

Wayne O. Chase

Abstract Words carry objective or *denotative* meanings that are agreed upon by a community and are held in the common mind of the community. But words and other forms of communication such as images, music, film, and architecture carry *connotative* meanings as well. These connotative meanings – mainly emotional associations – are also agreed upon by the community, and held in the common mind. However, access to a comprehensive storehouse of this enormous parallel universe of emotional meaning has never been available for the benefit of individuals, researchers, and businesses, in part due to the traditional separation of emotion and cognition in scientific research. Connotative intelligence technology, a system for capturing, quantifying, and making available the connotative meanings of words, images, music, and other artefacts of human culture and communication, is now being implemented in commercial products.

7.1 The Mother of All Technologies

If one were to confer upon a single technology the title, "mother of all technologies," surely it would be language. As a species, humans have been creators and users of the technology of oral language for perhaps 100,000–200,000 years. We have been users of the technology of written language for at least 5,000 years (Pinker, 1994, 1999; Dawkins, 2004). Good evidence exists that humans were creating and using elementary written language more than 30,000 years ago (Lumsden and Wilson, 1983).

The language functionality of the human brain may be thought of as an "experience lab" (Boekhorst, 2008). Every day, we make heavy use of our personal language experience lab. We address memory of past life experience, summon memory of words and their definitions, and employ grammatical functionality, in order to assemble complex information and exchange it with others. Without this skull-based lab, humans would exist merely at the level of other great apes such as chimpanzees, gorillas, and orangutans.

W.O. Chase (✉)
Connotative Intelligence Corporation, Vancouver, BC, Canada
e-mail: wchase@connotative.com

J. Westerink et al. (eds.), *Sensing Emotions*, Philips Research Book Series 12,
DOI 10.1007/978-90-481-3258-4_7, © Springer Science+Business Media B.V. 2011

Unlike our great ape cousins, humans have evolved areas of the brain specialized for language learning (Chomsky, 1972; Pinker, 1994, 1997; Jackendoff, 1994, 2002), likely thanks in large part to a mutation of the forkhead box P2 or FOXP2 gene (Pinker, 2001; Lai et al., 2001; Enard et al., 2002). Lacking the human version of FOXP2, chimpanzees, for example, cannot articulate speech, and are only capable of lifetime learning of around two hundred sign-language word-symbols. Without brain modules for language, their skills and language comprehension do not progress beyond those of a human 2-year-old (Wilson, 1978; Pinker, 1997).

As for written language, without it, humans would subsist only as hunters, gatherers, and primitive agriculturists, as our forebears did thousands of years ago, before the advent of literacy.

7.2 "Two Vocabularies Using the Same Set of Words"

Over the past few centuries, we have developed certain tools to help us make better use of the technology of language,. The two main tools are, of course, the dictionary and thesaurus. But these tools provide us with access to only half of the meaning that a word represents. Each word in a language actually encodes two distinctly different types of meaning, namely, denotative and connotative. As the philosopher Richard M. Weaver put it, denotation, or description, and connotation, chiefly feelings, "represent two vocabularies using the same set of words" (Weaver, 1974).

While we have had access to storehouses of the denotative, or descriptive, vocabulary for hundreds of years in the form of dictionaries and thesauruses, we have never had access to storehouses of the connotative, or feeling, vocabulary.

Here is a typical *Oxford English Dictionary* denotative definition: "*violin:* a musical instrument with four strings of treble pitch played with a bow." That definition is fine as far as it goes. But how do people *feel* about a violin?

Like denotative meaning, connotative meaning is "owned by the community" (Pinker, 2007). That is, people who share a language agree not only on the denotative meanings of words, but also on their connotative meanings. The difference is that the denotative definition *is* available in a dictionary or thesaurus, but the connotative definition *is not* – at least not yet. There are several reasons for this.

7.3 Past Barriers to Creation of a Connotative Dictionary

Until comparatively recently, a number of barriers have stood in the way of capturing connotative definitions and storing them in databases.

First, *adequate psychometric tools* needed to be developed to accurately quantify feelings and attitudes. This advancement did not take place until the first half of the twentieth century, when it was successfully addressed by psychometric specialists such as Guttman, Thurstone, and Likert (Carroll, 1960; Kidder, 1981; Uebersax, 2006).

Second, the *underlying dimensions of connotative meaning* – that is, the components of the feelings all humans have in common about objects and concepts, and the connection of those feelings to words and phrases – needed to be identified and described. (In the traditional separation of emotion and cognition in scientific research, language has usually been treated as an area of cognition.) This problem was solved by Osgood and colleagues at the University of Illinois in the 1950s (Osgood, 1952; Osgood et al., 1957).

Third, *comprehensive database-building systems* needed to be worked out that would capture connotative information in databases. Such databases would link the entire spectrum of emotional valence and intensity to the full range of commonly-used words of a language. This was the contribution of the author, a task begun in the early 1980s, culminating in the early 2000s with the granting of a family of five patents, the connotative intelligence patents.

Fourth, *personal computers needed to improve*. Connotative databases, and the products that would flow from them, would not easily lend themselves to print embodiments. Therefore it was essential, for the successful marketing of connotative products, that personal computers be widely available with sufficient memory, processing power, and operating system stability, to easily handle huge, graphics-rich databases speedily, and without freezing or crashing. Such capacity became commonly available in low-cost personal computers around 2003.

Lastly, an *adequate level of financing* was required to initiate commercial development of connotative databases and connotative language reference consumer products. In 2008, sufficient funding became available, and a handful of small Canadian companies began developing the first connotative language reference tools.

7.4 Osgood and the Discovery of E-P-A

A key figure in clarifying and advancing our understanding of connotative meaning was Charles Osgood, who devised an attitude scale called the semantic differential. *Semantics* refers to the meanings of words, and *differential* to a type of rating scale that records, along a continuum, a person's attitude towards an object or concept. A respondent, presented with a concept such as "violin," indicates his or her attitude by choosing a point along a continuum anchored at each end by antonyms such as "good-bad," "soft-hard," "weak-strong," etc. – pairs of polar opposites (Osgood, 1952; Osgood et al., 1957). With data from enough subjects and enough semantic differential scales, it was possible to identify, using factor analysis, a small number of underlying dimensions of connotative meaning.

In his investigations, Osgood used *Roget's Thesaurus* (Osgood et al., 1957; Griffin, 1991) to identify 289 word pairs having polar-opposite denotative meanings. But because of limitations of the University of Illinois' ILLIAC computer (this was the 1950s), Osgood's team had to trim the number to 76 pairs. (Osgood et al., 1957). Nevertheless, Osgood and colleagues were able to clearly identify three major underlying dimensions of connotative meaning, which Osgood termed *evaluation*, or E, *potency*, or P, and *activity*, or A (Osgood et al., 1957).

Over more than half a century, Osgood's findings have been validated thousands of times. Regardless of language or culture, the same three dimensions of connotative meaning invariably emerge. Today, the E-P-A structure of connotative meaning is considered one of the most thoroughly validated findings in social psychology (Osgood, 1969; Szalay and Bryson, 1974; Osgood et al., 1975; Tzeng, 1975; Chapman et al., 1980; Kidder, 1981; Griffin, 1991; Heise, 1969, 1970, 1992; Heise and Calhan, 1995; Bainbridge, 1994; Brewer, 2004).

The evaluation dimension is the affective component of connotative meaning, and the most important of the three – so much so that the term "connotative meaning" commonly refers mainly to the affective or emotional associations of a word or phrase (Maguire, 1973; Jerome, 1979; McArthur, 1992; Carroll, 1995; Crystal, 1995). In test after test, evaluation has accounted for most the variance. Potency and activity, while important, account for much less of the variance (Osgood, 1971; Oskamp, 1977; Brewer, 2004). A few other factors also emerge, but tend not to be nearly as prominent as E, P, and A – especially E (Osgood et al., 1957; Griffin, 1991; Bainbridge, 1994).

Semantic atlases have been compiled for research purposes. These are mini-connotative dictionaries that provide connotative profiles of 300–1,500 words (Jenkins et al., 1958; Heise, 1965; Snider and Osgood, 1969). A semantic atlas shows basic E-P-A ratings on each word, but does not provide connotative information on the broad spectrum of emotions subsumed by the evaluation dimension (Komorita and Bass, 1967).

7.5 E-P-A and Darwinian Natural Selection

Since evaluation, potency, and activity emerge as the dominant dimensions of connotative meaning in cultures and languages worldwide, it is reasonable to hypothesize, as Osgood did, that the universality of the E-P-A assessment capacity in humans is rooted in conventional Darwinian natural selection (Osgood, 1969, 1971; Chapman et al., 1980; Griffin, 1991; Bainbridge, 1994), especially considering that emotions themselves are evolutionary adaptations (Darwin, 1872/1998; Pinker, 1997).

Hundreds of thousands of years ago, an individual, when confronted with something unexpected, such as a saber-tooth tiger, a rabbit, a forest fire, an attractive person, a clap of thunder, or an unfamiliar tribesman, would have had to make a quick but accurate assessment of the unexpected thing. The *evaluation* (emotional) response would dominate. In an instant, the individual would experience an emotion-driven reaction: to fight, to flee, to take delight in, etc. Simultaneously, there would be an instantaneous *potency* assessment: is this thing bigger and more powerful than me, or smaller and weaker? And, at the same time, an immediate *activity* judgment: is this thing active and fast, or is it slow, or is it totally inactive?

This quick, automatic E-P-A assessment would at times spell the difference between life and death. Survival success meant that E-P-A brain circuitry would be passed on genetically and eventually become encoded (with evolving language functionality) in the meanings of words.

7.6 Connotative Intelligence

Moving from history to the present day, as mentioned, practical systems for capturing, in a database, the connotative content of an entire language have now been developed and patented. These systems make it possible, for the first time, to create connotative dictionaries, connotative thesauruses, connotation-checkers, and other connotative language reference tools.

The methodology calls for the construction a series of rating scales to measure connotative meaning in an absolute, context-independent way, using discrete visual analog scales (Uebersax, 2006). The author's research findings, which reveal very high correlations between individual raters' scores and group averages, support this approach, as do the findings of other investigators (Jenkins et al., 1958; Ware et al., 1970; Mehrabian, 1990, 1997, 2001).

Details of the five Connotative Intelligence patents are beyond the scope of this paper, but the patents are a matter of public record and available at online patent servers such as Google's. The patent titles are:

System for Identifying Connotative Meaning
System for Quantifying Intensity of Connotative Meaning
Interactive Connotative Dictionary System
Interactive Connotative Thesaurus System
System for Connotative Analysis of Discourse

7.7 Overview of Connotative Language Reference Products

Development of connotative language reference tools is now underway. These new tools will be both fascinating and fun to use. Here are some details of what a user of connotative language tools in everyday life will see on his or her computer screen.

Connotative Dictionary. The connotative dictionary will have several major characteristics that will distinguish it from its familiar denotative cousin.

First, unlike the denotative definition of a word, the connotative definition, or *connotative profile*, will take the form of a *graphic image*, with minimal text. The bars on the graph will represent quantified emotional valences and intensities, as well as intensities of potency and activity. Some text will accompany the image to identify the word and context, but the main component of a connotative profile will be a graphic image (Fig. 7.1).

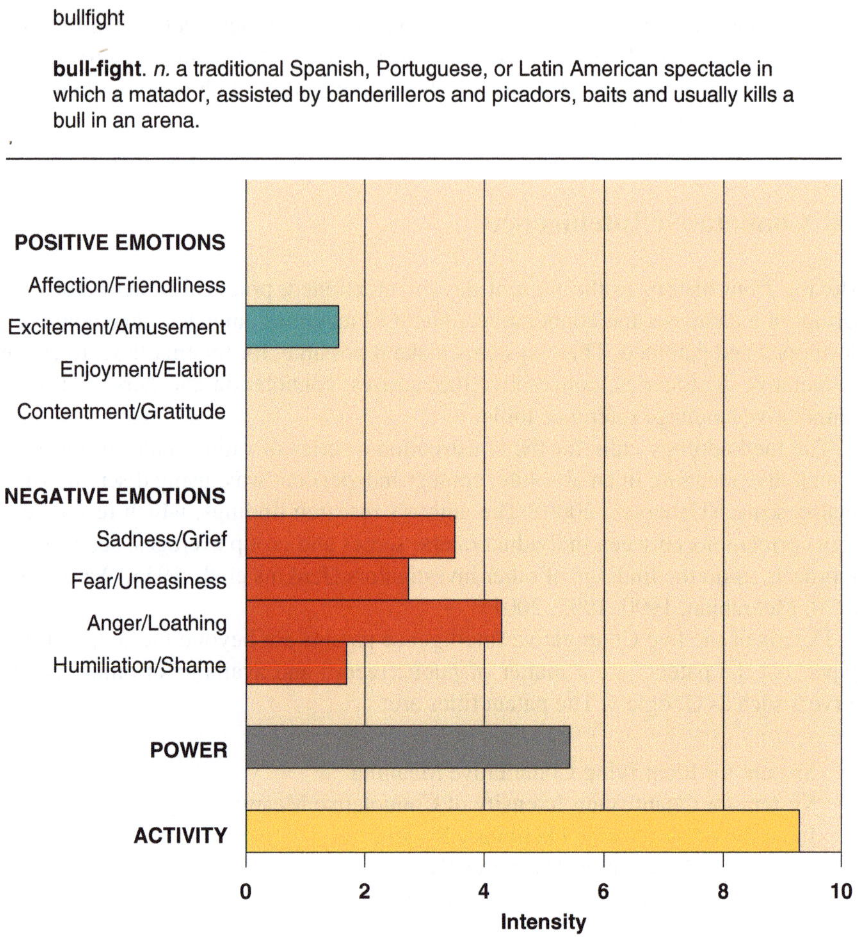

Fig. 7.1 A typical connotative dictionary entry

Second, because of the space required to display a graph and its labels, it is unlikely that a connotative dictionary of the English language or any language will ever appear in print. Even if as many as 12 small graphs could be printed on a page, a printed volume of 1,000 pages would hold only 12,000 connotative profiles – not nearly enough to adequately represent the commonly-used words of a whole language. Therefore, a complete connotative dictionary will necessarily be a digital product.

Third, a connotative dictionary will include a large proportion of *proper nouns and terms from popular culture*, such as first names of people, names of well-known cities, products, corporations, and celebrities from all walks of life, as well as a variety of slang terms, idioms, and even catch phrases from the advertising industry. These entities have strong E-P-A associations.

"Philips," the corporation, for example, has a certain public image, and therefore the company name "Philips" will have its own connotative profile (actually a series of them – one for each nation and cultural group), which users will be able to compare with the connotative profiles of other corporate names, such as "Microsoft" or "Braun." The connotative profile of any given corporate name will vary considerably by nation and identifiable cultural group.

The English language has a 1,500 year history (McCrum et al., 1986; Crystal, 1995; Winchester, 2003), and the Oxford English Dictionary defines upwards of half a million words (McCrum et al., 1986; Winchester, 2003). The first English language connotative dictionary will be applicable only within a defined geographical region, and will likely have in the order 50,000 connotative profiles. These will be the words most widely known and most commonly used by most people within the region. Most English speakers have a total vocabulary that ranges from about 30,000 to 100,000 words (McCrum et al., 1986; McArthur, 1992; Pinker, 1994).

The connotative dictionary will provide teachers and learners of a second with access not only to the dictionary meanings of all the important words in the second language, but also to the full spectrum of connotative meanings of those words.

Connotative Thesaurus. The user of a connotative thesaurus will be able to call up the connotative profile of any word, then find other words that match that connotative profile. The resulting list of words will be *connonyms* – connotative synonyms. They will be related to each other by the similarity of emotional valence and intensity that they elicit in the population, but will have entirely different denotative meanings.

Connonyms will be especially useful in the creation of metaphors. Metaphor pervades vivid writing and enables a language to grow by extending new meanings to words and phrases. New metaphors grab and hold attention because they are unexpected (Heath and Heath, 2007). Metaphor in one form or another has always been the primary means of emotional expression in great writing (Brooks and Warren, 1958; Jerome, 1979; Brewer, 2004).

The connotative thesaurus will, in effect, be a thesaurus of highly accurate and emotion-eliciting metaphors.

Connotative Language Translator. Once connotative databases are available in multiple languages, it will be possible to incorporate connotative meaning into automated language translation, which should improve the emotional "feel" of the translation. The software will not improve syntax, but the overall accuracy of the translated message should improve.

Connotation Checker (emotional tone checker). From a functional standpoint, this application will work something like a grammar-checker. It will scan a passage of text and report on the text's emotional tone. The user will then be able to completely change the emotional valence and emotional intensity of the original passage (and thus, the emotional effect on the user's intended audience) by removing certain words and replacing them with words having connotative profiles with different emotional valences and quantified intensities. The application will advise the writer on the use of suggested words and phrases, so that the writer will be able to significantly change the emotional tone without drastically changing the subject matter of the message. Unlike a grammar checker, a connotation checker will not attempt to automatically re-write the text; the user will retain control of syntax (Kies, 2008).

Anyone writing in everyday contexts, be it an email, blog entry, tweet, essay, speech, novel, news release, or corporate memorandum will have control over both the objective content and the emotional impact of their text.

Writing that tends to be compelling and memorable contains highly charged vocabulary. The connotation checker will enable the user to flag and remove dull, low-E-P-A words, and replace them with more emotionally loaded vocabulary. The best, most memorable writing does not merely convey information, but persuades and entertains, so as to hold the attention of the reader or listener (Weaver, 1974; Jerome, 1979; Bestgen, 1994). Such writing includes television and movie dialogue, song lyrics, novels and short stories, poetry, humour, reviews, political and religious writing, advertising and PR writing, sports writing, travel writing, speech writing, editorial commentary, blogging, tweeting, and related social media writing. What all of these varieties of writing have in common – in other words, what the great majority of effective writing has in common – is that the deliberate use of intense connotative meaning plays a central role.

The connotation checker is least likely to benefit academics and technologists, whose job is to communicate, as far a possible, objective, unbiased information. Academics seek to minimize vocabulary loaded with strong connotative associations. In so doing, academic and technical writing must necessarily break all rules of compelling, memorable communication (alas, that includes the chapter you are now reading!). Academic and technical writing typically employs the passive voice, long and complex sentences, few or no personal pronouns, few non-declarative sentences, and a style devoid of narrative. The result is boring, forgettable writing, *except* to other academics and technologists. The connotation checker might only be of use to such writers as a means of seeking out and removing any vocabulary that has a connotative pulse.

The connotation checker may also provide indexes such as these:

- An *abstract usage* index. The hundred-thousand-year-old brains of humans much prefer *concrete* vocabulary (words that appeal to the five senses) over abstract vocabulary (Flesch, 1949; Brooks and Warren, 1958; Godfrey and Natalicio, 1970; Jerome, 1979; Heath and Heath, 2007). The connotation checker may provide the user with an index of abstract usage and warn the user as abstract vocabulary increases. For academic and technical writing, the abstract usage index will be off the charts; such writing tends to be overwhelmingly abstract.
- An index of usage of *personal words*, as originally defined by Rudolph Flesch (1949); the more, the better.
- An index of usage of *personal sentences*, as originally defined by Flesch (1949); again, the more, the better.
- An index of usage of *"Hayakawa" words*, an indicator of clarity in writing (Hayakawa and Ehrlich, 1994).

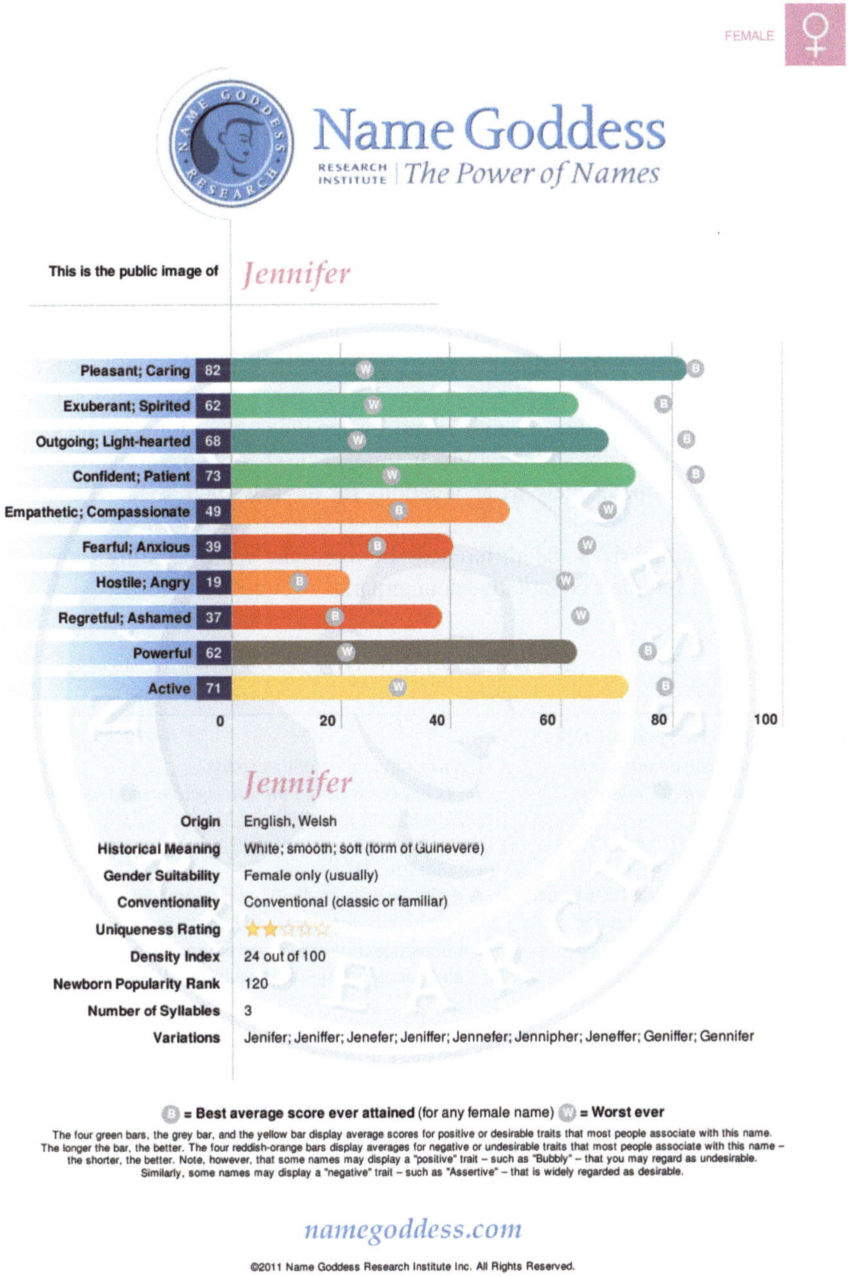

Fig. 7.2 A typical public image profile

7.7.1 Other Products Based on Connotative Profiles

He meant/she meant (or "Mars and Venus") connotationary and connosaurus. These variants would shows differences between men's and women's connotative profiles for the same word.

Connotative person namer. This variant (now in commercial development) will provide connotative profile data, called *public image profiles*, on thousands of first names (Fig. 7.2).

Connotative business and product namer. Similar to the person-naming connotative products now in development, but incorporating a full-language database.

Connotative image database, song database, movie database, product database, etc. These databases will enable the user will to pre-select emotional valence and intensity, then search for images, songs, movies, etc. that match the user's selection.

Connotative word games. Games based on language have long been popular, such as crossword puzzles and Scrabble. In time, games based on connotative meanings will be developed, likely as applications for devices such as the iPhone and Blackberry.

Table 7.1 summarizes the differences between the main existing denotative language tools, and their connotative counterparts.

Table 7.1 Denotative language tools and their connotative counterparts

Existing denotative language tools *Tools to access the denotative universe of meaning ("What is it?")*	Future connotative counterparts *Tools to access the connotative universe of meaning ("How does it feel?")*
Dictionary A database of denotative definitions of each word in the language	*Connotative Dictionary* A database of both denotative definitions and graphically-represented connotative valences and intensities of widely-used English language words; also incorporating idioms and widely-known proper nouns
Thesaurus A database of synonyms – words grouped by similarity of denotative meaning (ideas)	*Connotative Thesaurus* A database of "connonyms" – words grouped by similarity of connotative meaning (specific emotions evoked); a group of connonyms will rarely incorporate any terms that are synonymous
Grammar and Spell Checkers Software that checks the accuracy of grammar and spelling	*Connotation Checker* Software that reports on the overall emotional tone (and other connotations) of a text passage and provides the user with alternative words and phrases to change the tone
Electronic Denotative Language Translators Software that automatically translates text across languages, according to denotative meaning	*Electronic Denotative/Connotative Language Translators* Software that automatically translates text across languages, according to both denotative and connotative meaning

7.8 Connotative Products in Development

As of 2009, several companies were involved in bringing connotative language products to market. One project is the connotative language checker, as described above. Another is an application that analyses the emotional tone of Twitter tweets. A third is a combined connotative dictionary and connotative thesaurus of the first names of natural persons. Mehrabian published the first connotative dictionary of first names in 1990. Although it was limited in scope to connotative profiles comprised of only half a dozen scales associated with each name, it was nonetheless commercially successful, demonstrating the viability of a connotative product in the marketplace (Lawson, 1971; Mehrabian, 1990, 1997, 2001).

7.9 Accessing the Parallel Universe of Connotative Meaning: What's Next?

Language being the "mother of all technologies," it is not surprising that there has long been a huge proven market for language tools focused on word meanings:

Dictionary. Dictionaries have been around for hundreds of years and are perennial best-sellers. They are so useful that they are built into word processing applications such as Microsoft Word and WordPerfect. According to organizations such as Quantcast.com and Compete.com, which publish website traffic data, the website *Reference.com*, which provides free lookup of words in online dictionaries and thesauruses, averages more than 10 million unique US visitors monthly, and is one of the top 100 most-visited websites.

Thesaurus. When the first comprehensive thesaurus was introduced in 1852, the public immediately grasped its usefulness. *Roget's Thesaurus* was an immediate hit, reprinted 28 times in various editions before Roget died 17 years later (Atkinson, 2001; Kendall, 2008). It is fitting that today, the word "Roget's" is itself a synonym for thesaurus. Like the dictionary, the thesaurus is also built into Microsoft Word and WordPerfect.

Connotative language reference tools. Since *Roget's Thesaurus* was published, there have been no new language tools that provide new information on word meanings. Connotative language reference tools will be the first language reference tools since 1852 to make available new aspects of meaning (connotative profiles) associated with the majority of words and phrases in everyday use by the majority of speakers of a language.

Connotative meaning is a universe of language meaning that parallels the universe of denotative meaning. Everyone has had access to the denotative universe for generations. Connotative language tools will finally provide access to the connotative universe.

References

Atkinson, W.I.: Prototype: How Canadian Innovation Is Shaping the Future. Thomas Allen, Toronto (2001)

Bainbridge, W.S.: Semantic differential. In: Asher, R.E., Simpson, J.M.Y. (eds.) The Encyclopedia of Language and Linguistics, vol. 7. Pergamon Press, Oxford, 3800–3801 (1994)

Bestgen, Y.: Can emotional valence in stories be determined from words? Cogn. Emot. **8**, 21–36 (1994)

Boekhorst, F.: Foreword. In: Westerink, J.H.D.M., Ouwerkerk, M., Overbeek, T.J.M., Pasveer, W.F., de Ruyter, B. (eds.) Probing Experience: From Assessment of User Emotions and Behaviour to Development of Products. Springer, Dordrecht, The Netherlands, vii–viii (2008)

Brewer, W.F.: Charles Osgood: The psychology of language. In: Hoddeson, L. (ed.) No Boundaries: University of Illinois Vignettes. University of Illinois Press, Urbana, 210–225 (2004)

Brooks, C., Warren, R.P.: Metaphor. In Cleanth Brooks and Robert Penn W. (ed.): Modern Rhetoric, 2nd edn. Harcourt, Brace & World, New York, NY (1958)

Carroll, J.B.: Vectors of prose style. In: Seboek, T.A. (ed.) Style in Language. Wiley, New York, NY, 593–602 (1960)

Carroll, D.A.: A Manual of Writers' Tricks, 2nd edn. Marlowe, New York, NY (1995)

Chapman, R.M., McCrary, J.W., Chapman, J.A., Martin, J.K.: Behavioral and neural analyses of connotative meaning: word classes and rating scales. Brain Lang. **11**, 319–339 (1980)

Chomsky, N.: Language and Mind, Enl. edn. Harcourt Brace Jovanovich, New York, NY (1972)

Crystal, D.: The Cambridge Encyclopedia of the English Language. Cambridge University Press, Cambridge (1995)

Darwin, C. The expression of the emotions in man and animals. 3rd edn. Oxford University Press, New York, NY (1872/1998)

Dawkins, R.: The Ancestor's Tale: A Pilgrimage to the Dawn of Life. Weidenfeld & Nicolson, London (2004)

Enard, W., Przeworski, M., Fisher, S.E., Lai, C.S.L., Wiebe, V., Kitano, T., Monaco, A.P., Paabo, S.: Molecular evolution of FOXP2, a gene involved in speech and language. Nature **418**, 869–872 (2002)

Flesch, R.: The Art of Readable Writing. Harper & Row, New York, NY (1949)

Godfrey, R.R., Natalicio, L.F.S.: Evaluation on semantic differential equals abstraction plus error. Psychol. Rep. **27**, 467–473 (1970)

Griffin, E.: Mediational theory of meaning of Charles Osgood. In E.M. Griffin (ed.): A First Look at Communication Theory. McGraw-Hill, New York, NY, 28–37 (1991)

Hayakawa, S.I., Ehrlich, E.: Choose the Right Word: A Contemporary Guide to Selecting the Precise Word for Every Situation. HarperCollins, New York, NY (1994)

Heath, C., Heath, D.: Made to Stick: Why Some Ideas Survive and Others Die. Random House, New York, NY (2007)

Heise, D.R.: Semantic differential profiles for 1,000 most frequent English words. Psycholog. Monogr. **79**(whole no. 601) (1965)

Heise, D.R.: Some methodological issues in semantic differential research. Psycholog. Bull. **72**, 406–422 (1969)

Heise, D.R.: The semantic differential and attitude research. In: Summers, G.F. (ed.) Attitude Measurement. Rand McNally, Chicago, 235–253 (1970)

Heise, D.R.: Affect control theory and impression formation. In: Borgatta, E., Borgatta, M. (eds.) Encyclopedia of Sociology, vol. 1. Macmillan, New York, NY, 12–17 (1992)

Heise, D.R.: Project Magellan: Collecting cross-cultural meanings via the internet. Electron. J. Sociol. **5** (2001). http://www.sociology.org/content/vol005.003/mag.html. Accessed 12 Sept 2008

Heise, D.R., Calhan, C.: Emotion norms in interpersonal events. Soc. Psychol. Q. **58**, 223–240 (1995)

Jackendoff, R.: Patterns in the Mind: Language and Human Nature. Basic Books, New York, NY (1994)
Jackendoff, R.: Foundations of Language: Brain, Meaning, Grammar, Evolution. Oxford University Press, Oxford (2002)
Jenkins, J.J., Wallace, A.R., Suci, G.J.: An atlas of semantic differential profiles for 360 words. Am. J. Psychol. **71**, 688–699 (1958)
Jerome, J.: The Poet and the Poem. Writer's Digest Books, Cincinnati (1979)
Kendall, J.: The Man Who Made Lists: Love, Death, Madness, and the Creation of Roget's Theaurus. Putnam, New York, NY (2008)
Kidder, L.H.: Research Methods in Social Relations, 4th edn. Holt, Rinehart & Winston, New York, NY (1981)
Kies, D.: Evaluating grammar checkers: a comparative ten-year study. Hypertext Books. http://papyr.com/hypertextbooks/grammar/gramchek.htm (2008). Accessed 26 May 2008
Komorita, S.S., Bass, A.R.: Attitude differentiation and evaluative scales of the semantic differential. J. Personal. Soc. Psychol. **6**, 241–244 (1967)
Lai, C.S.L., Fisher, S.E., Hurst, J.A., Vargha-Khadem, F., Monaco, A.P.: A forkhead-domain gene is mutated in a severe speech and language disorder. Nature **413**, 519–523 (2001)
Lawson, E.D.: Semantic differential analysis of men's first names. J. Psychol. **78**, 228–240 (1971)
Lumsden, C.J., Wilson, E.O.: Promethean Fire: Reflections on the Origin of Mind. Harvard University Press, Cambridge (1983)
Maguire, T.O.: Semantic differential methodology for the structuring of attitudes. Am. Educ. Res. J. **10**, 295–306 (1973)
McArthur, T. (ed.): The Oxford Companion to the English Language. Oxford University Press, New York, NY (1992)
McCrum, R., Cran, W., MacNeil, R.: The Story of English. Viking Penguin, New York, NY (1986)
Mehrabian, A.: The Name Game: The Decision That Lasts a Lifetime. National Press Books, Bethesda (1990)
Mehrabian, A.: The revised name connotation profile. Available from author (1997)
Mehrabian, A.: Characteristics attributed to individuals on the basis of their first names. Genet. Soc. Gen. Psychol. Monogr. **127**, 59–88 (2001)
Osgood, C.E.: The nature and measurement of meaning. Psychol. Bull. **49**, 197–237 (1952)
Osgood, C.E.: On the whys and wherefores of E, P, and A. J. Personal. Soc. Psychol. **12**, 194–199 (1969)
Osgood, C.E.: Exploration in semantic space: a personal diary. J. Soc. Issues **27**, 5–64 (1971)
Osgood, C.E., May, W.H., Miron, M.S.: Cross-Cultural Universals of Affective Meaning. University of Illinois Press, Urbana, IL (1975)
Osgood, C.E., Suci, G.J., Tannenbaum, P.H.: The Measurement of Meaning. University of Illinois Press, Urbana, IL (1957)
Oskamp, S.: Attitudes and Opinions. Prentice-Hall, Englewood Cliffs, NJ (1977)
Pinker, S.: The Language Instinct. HarperCollins, New York, NY (1994)
Pinker, S.: How the Mind Works. W. W. Norton & Co, New York, NY (1997)
Pinker, S.: Words and Rules: The Ingredients of Language. Basic Books, New York, NY (1999)
Pinker, S.: Talk of genetics and vice versa. Nature **413**, 465–466 (2001)
Pinker, S.: The Stuff of Thought: Language as a Window into Human Nature. Penguin, New York, NY (2007)
Snider, J.G., Osgood, C.E. (eds.): Semantic atlas for 550 concepts. In: Semantic Differential Technique: A Sourcebook. Aldine, Chicago, 625–636 (1969)
Szalay, L.B., Bryson, J.: Psychological meaning: Comparative analyses and theoretical implications. J. Personal. Soc. Psychol. **30**, 860–870 (1974)
Tzeng, O.C.S.: Reliability and validity of semantic differential E-P-A markers for an American english representative sample. Psychol. Rep. **37**, 292 (1975)
Uebersax, J.S.: Likert scales: dispelling the confusion. Statistical methods for rater agreement, http://ourworld.compuserve.com/homepages/jsuebersax/likert.htm (2006). Accessed 9 Sept 2008

Ware, E.E., Radtke, R.C., Berner, E.: The effect of context on semantic differential ratings. Can. J. Behav. Sci. **2**, 246–252 (1970)
Weaver, R.M.: A Rhetoric and Composition Handbook. William Morrow, New York, NY (1974)
Wilson, E.O.: On Human Nature. Harvard University Press, Cambridge (1978)
Winchester, S.: The Meaning of Everything: The Story of the Oxford English Dictionary. Oxford University Press, Oxford (2003)

Chapter 8
Runners' Experience of Implicit Coaching Through Music

Joyce H.D.M. Westerink, Arjan Claassen, Tijn Schuurmans, Wijnand IJsselsteijn, Yvonne de Kort, Suzanne Vossen, Yuzhong Lin, and Guido van Helvoort

Abstract In this paper we evaluate a music-based coaching system for runners, the SportsCoach. It measures the runner's heart rate and increases music tempo when, for an optimal workout, the runner should speed up. Coaching is implicit, since the runner only needs to keep in sync with the music and no explicit instructions are given. We performed 2 experiments to evaluate how this implicit coaching was experienced in the actual context of running. The first experiment investigated how natural it is to keep running in sync with the music when the music tempo changes. We find that although runners are not naturally inclined to do so, a band of 10% below one's natural tempo is mostly easily followed, especially by dancers. The second experiment evaluated the SportsCoach and contrasted its implicit form of coaching and synchronized music to explicit and absent forms of coaching and fixed tempo music. We find that the SportCoach concept scores well on most aspects, especially because of the synchronicity of music and running tempos.

8.1 Introduction

8.1.1 Coaching and Music in Sports

Many people exercise. Sometimes because they like doing so, sometimes because they pursue another goal, for instance losing weight. The former group is intrinsically motivated (Deci and Ryan, 1985), and might even experience "flow", a concept introduced by Csikzentmihalyi (1990) denoting a perceived optimal balance between skills and challenges. Other people (the latter group) are extrinsically motivated, and especially for them there is a discrepancy between the task of exercising and the goal they have set themselves. Sometimes the presence of a coach can help. Westerink et al. (2004) and Ijsselsteijn et al. (2006) describe how virtual coaches (and virtual realities) can help raise the motivation of the people who exercise. In

J.H.D.M. Westerink (✉)
Brain, Body and Behavior Group, Philips Research, Eindhoven, The Netherlands
e-mail: joyce.westerink@philips.com

J. Westerink et al. (eds.), *Sensing Emotions*, Philips Research Book Series 12,
DOI 10.1007/978-90-481-3258-4_8, © Springer Science+Business Media B.V. 2011

general, they found that explicit instructions (e.g. "your heart rate is too high. Please slow down") are less motivating than more implicit guidance, for instance when the advice is offered in terms of a graph. In this paper we investigate the possibility of giving implicit guidance through music.

Many people already exercise to music. It is for instance common for joggers to use music as a distraction. But it also can provide stimulation that can help to perform better. In an article summarizing various studies on the effects of music in a sports setting Karageorghis and Terry (1997) state that there are inconsistent findings, but music does seem to have a positive effect in sub-maximal exercise. Anshel and Marisi (1978) state that "music, particularly if synchronised to physical movement, has a positive effect on the ability to endure the task". Boutcher and Trenske (1990) found that for low and moderate workouts, music leads to lower perceived exertion than a simple metronome does. Tenenbaum et al. (2004) observed that music was perceived as beneficial by many, since it directed attention to the music and motivated to continue running.

We developed the SportsCoach to combine all the positive effects of music described above: raised motivation and lower perceived exertion, synchronization to physical movements as well as the advantages of implicit guidance.

8.1.2 The SportsCoach

The conceptual design and architecture of the SportsCoach has been described by Wijnalda et al. (2005). Also others have introduced similar concepts, e.g. Oliver and Kreger-Stickles (2006). In this paper we therefore confine the description of the SportsCoach to what is needed for understanding the experiments to be presented.

The SportsCoach was implemented on an iPaq PDA, on which a number of contemporary popular songs with regular, constant tempos was compiled. The collection covered a wide range of tempos. Music playback software ensured that the tempos of these songs could be raised or lowered within a range of 25% of the original tempo, without noticeable negative effects in music quality.

As input, the SportsCoach received the runner's heart rate signals from a dedicated chest belt. If the heart rate was higher than required for the workout schedule, the SportsCoach would lower the tempo of the music. Thus a runner keeping in pace with the music would automatically slow down, which in turn was expected to lower the heart rate. Vise versa, for a heart rate that was measured to be too low, music tempo would be raised in order to increase workout level and trough it heart rate.

A second input to the SportsCoach was a measurement of the runner's running tempo by means of a timed step counter. This input is needed to correct for possible discrepancies between music tempo and runner's tempo. Also, it allowed an additional feature of the SportsCoach: the possibility for the user to select the so-called "Follow" mode, in which the music tempo simply adapts to the running tempo, and no (implicit) coaching is given.

8.1.3 Research Questions

Military parades are often put forward to prove humans' innate tendency to adapt their repetitive movements to the tempo of music. Also Large (2000) describes an "often completely spontaneous musical synchronization". We were interested to see whether this synchronization also appears in our SportsCoach in the context of running, and how large the changes in music tempo can be for this to occur. Should natural synchronization not occur at all, the SportsCoach concept is still valuable if the runner at least is able to willingly follow the tempo changes in the music. In that case, it is necessary for the design of the SportsCoach to know which music changes are easily followed by the runners. Thus our first experiment focuses on the naturalness and ease of following music tempo changes while running.

Whether completely natural or just very easy, the SportsCoach is intended to raise runners' motivation and enhance their enjoyment in running. Two factors are mainly expected to account for this: the implicit form of coaching and the synchronicity of movements to music. So the main intention of our second experiment is to see whether the SportsCoach indeed succeeds in raising motivation, lowering perceived exertion and perhaps attaining a state of flow. In addition, we wanted to investigate the relative contributions of the factors implicit coaching and synchronicity to the effects found.

8.2 Experiment 1: Ease and Naturalness of Following the Music Tempo

8.2.1 Design

The experiment used a two factor within-subjects design. Independent variables were:

- Instructions (either with or without explicit instructions to run in sync to the music)
- Music tempo: This was defined as a percentage of the initial running tempo, and had 6 levels: 100, 97, 94, 91, 88, and 82%. The music tempos were presented in 1-min stages.

Dependent variables were derived from the continuously monitored running tempo:

- ρ, the average running tempo, calculated over the last 45 s of each stage of music,
- ε, defined as the average absolute deviation of the running tempo from the music tempo, averaged over the last 45 s of each music stage,
- τ, the time to adapt, defined as the duration between a music tempo change and the first instance the running tempo reaches a value within $\pm 2\%$ of it.

8.2.2 Participants

Twenty-three persons participated in the experiment (8 females), aged between 20 and 66 years (mean 35 years). They were selected for their stamina in sports, since the experiment required 40 min of running; some were members of an athletics group. They signed an informed consent form and received a reward of 15 € for participation.

8.2.3 Method

The experiment took place in a 20×40 m^2 sports hall. Participants first filled out a small questionnaire asking for demographics and their music skills. Then they were equipped with a headphone, a pedometer, and an early, laptop-based version of the SportsCoach, which was carried by the participants in a backpack, amounting to a total weight of approximately 4 kg. Thus the SportsCoach monitored both music tempo and running tempo throughout the experiment.

The participants did two separate sessions of 20 min each. Before each session they were given the opportunity for a warming-up with full equipment. Before the first session they were told they would hear music, but no mention was made of the music tempo changes that would take place during that session, nor were any explicit instructions given as to how they should run. Before the second session the participants' attention was drawn to the music changes they must have noticed during the first session, and now they were given the explicit instruction to try and run in sync to the music at all times.

In both sessions the procedure was the same. It started with a 3-min period in which the music adapted its tempo to the running tempo of the participant, allowing it to stabilize. This tempo was now taken as the initial running tempo. After these first 3 min, 17 one-minute stages with different music tempos were presented, in a fixed, predetermined order at respectively 100, 97, 94, 91, 100, 91, 94, 97, 100, 94, 88, 82, 100, 82, 88, 94, and 100% of the initial running tempo. We choose tempos below 100% since we assumed that in real-life running tempo decreases are more common than tempo increases.

8.2.4 Results and Discussion

Figure 8.1 shows the running tempo averaged over participants for each of the stages of both the "with" and the "without instruction" sessions, as well as the music tempo itself. From the figure it becomes clear, that without explicit instructions almost no running tempo changes were made by the participants, since the average remains steady at the initial running tempo level. The difference between with ($\rho_{\text{mean}} = 96.4\%$) and without instruction ($\rho_{\text{mean}} = 100.3\%$) also becomes apparent from a repeated measures ANOVA, which shows a significant effect of instruction on running tempo ($F(1,18) = 19.9$, $p < 0.01$). A second repeated measures ANOVA on the "without instruction" data only, reveals no significant effect of stage on running tempo ($F(2.7,51.3) = 1.83$, $p = 0.16$). Apparently, in the context of running with the SportsCoach runners are not naturally inclined to follow the rhythm of the music with their movements. In part, this might be due to the high percentage of trained runners in our sample, since they might have become accustomed to maintaining a steady rhythm during running. On the other hand, also the less-trained participants must have omitted to follow the music, since no significant changes in average running tempo were found, not even for the smaller changes in music tempo.

Figure 8.1 also shows that even in the "with instructions" condition runners do not follow the music tempo changes completely. This does not necessarily mean that none of the runners does so. Most probably there will be individual differences between runners. To investigate whether such individual differences were related to the musical skills of the runners, we conducted a third repeated measures ANOVA, this time on the ε-values of the "with instruction" sessions only, with stage as a within-subjects factor and musical skills (4 mutually exclusive levels: non-musicians, band-members, dancers, and soloists) as a between-subjects factor. We found a significant main effect of musical skills ($F(3,17) = 5.3$, $p < 0.01$). Apparently, not all runners always fully adapted their running tempo to the music tempo, not even in the second session, in which they had received explicit instructions to do so, as can be seen in Fig. 8.2. Band-members show very large ε-values, generally as big as the difference between music and initial running tempo, suggesting that although they did receive instructions to follow the music, basically none of them ever did. Dancers on the other hand appear to have the smallest ε-values, even for the stages 11 and 13, which have the lowest and most difficult music tempos, as can be seen from Fig. 8.2. In hindsight, it makes sense that it is not the musical training per se, that predicts one's ability to follow various running tempos, but rather one's skill in adapting one's bodily movement to music, as is trained through dancing.

As we have found that not everyone always is capable of adapting one's running tempo to the music tempo, it becomes interesting to investigate which tempo changes are easily followed and which are not. This is depicted in Fig. 8.3, which presents the percentage of runners that is able to adapt to a new music tempo. It appears to be mainly determined by the value of the new running tempo, expressed as a percentage of the initial tempo. The majority of runners (60%) can still successfully adapt to a music tempo that is 90% of their initial running tempo, whereas 90% of them can adapt to music tempos of minimally 97% of the initial running tempo. In the figure, several parameters indicate different sizes and directions of change, but all in all, they hardly influence how many runners are able to follow the changing music tempo. As for the time τ needed to adapt, we found considerable variation between instances, and calculated the average time needed to adapt as 11.2 s.

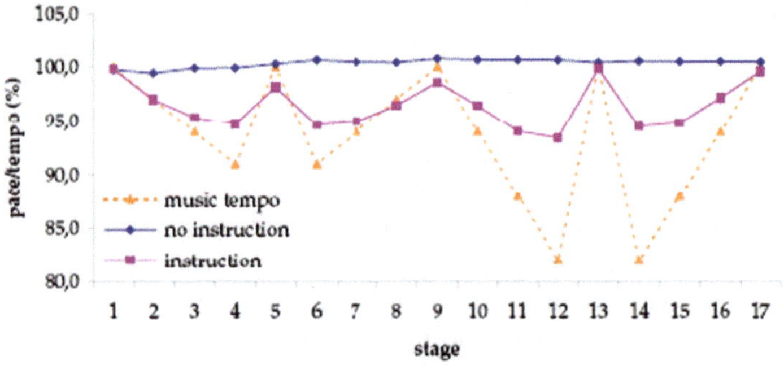

Fig. 8.1 Running tempo (*pace*) averaged over participants, as a percentage of the initial pace, per stage in the sessions with and without instructions, and contrasted to the music tempo

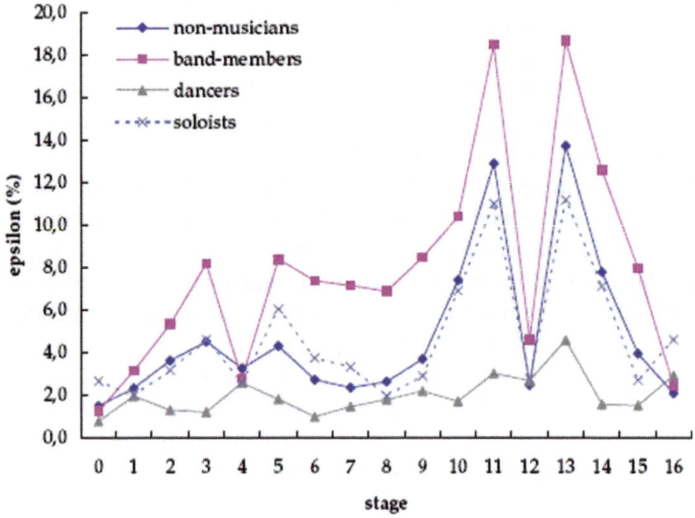

Fig. 8.2 Average ε-values of ε per stage and per group of runners with the same musical skills

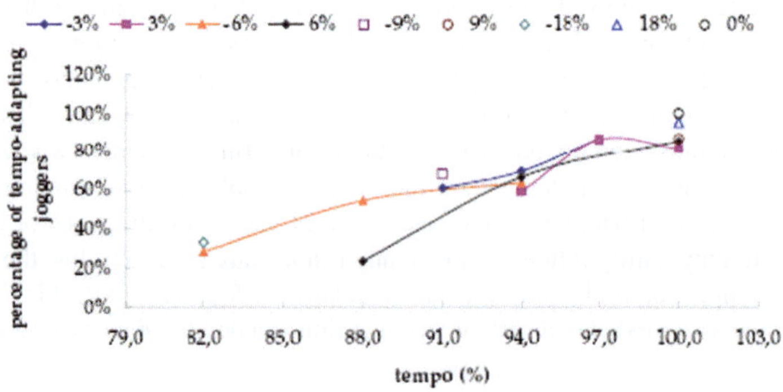

Fig. 8.3 The percentage of runners adapting to the music tempo as a function of the target running tempo expressed as a percentage of the initial running tempo. Different sizes and directions of change are indicated as parameters

8.2.5 Conclusion

This experiment showed that most runners can adapt their running tempos to music if they want, provided the tempo changes proposed are not too different from their initial running tempos. Thus the concept of the SportsCoach remains feasible, although it is unsure how natural and implicit the music-tempo changes really are as a coaching tool, since the runners did not appear to adapt unconsciously to the music tempo changes presented. To gain some insight in the user experiences with such synchronized music and possibly implicit coaching in the context of running, we set up the following experiment.

8.3 Experiment 2: User Experiences with Implicit, Synchronised Coaching Compared to Absent, Explicit And/Or Fixed-Music Coaching

8.3.1 Design

The experiment used a two factor within-subjects design. Independent variables were:

- Music mode, which had 2 levels: Running was either synchronous or asynchronous to the music.
- Type of coaching: Three levels were presented: No coaching or feedback at all, explicit feedback of heart rate (HR) as visible on a wristwatch, and implicit coaching through music.

Dependent variables focused on a variety of user-experience aspects as obtained through questionnaires[1]:

- Exertion was measured by means of the Borg Scale rating of perceived exertion (Chen et al., 2002),
- The occurrence of Flow during running was measured with the Flow Sate Scale 2 (FSS2) (Jackson and Eklund, 2002),
- Intrinsic motivation was measured by the Intrinsic Motivation Inventory (IMI)
- The motivational quality of music was rated in the Brunel Music Rating Inventory (MRI) (Karageorghis et al., 1999).

[1]Of course, we also obtained heart rate data in the logging file of the SportsCoach prototype. We will present those data in a future publication, however, since in this paper we want to focus on user experiences in the running context rather than on the functional performance of the SportsCoach.

8.3.2 Participants

Twelve persons participated in the experiment, mostly students under the age of 30. Since the previous experiment had indicated that professional and amateur runners might find difficulty in stepping out of their usual running rhythm, we deliberately selected participants with no professional or amateur running background for this experiment. They did all indicate they were in good enough shape to run the sessions. They signed an informed consent form and were rewarded with a sports radio for participation.

8.3.3 Method

The experiment took place outdoors in good running weather. The participants were equipped with a headphone, a chest-belt measuring heart rate and running pace, and an improved, iPaq-based version of the SportsCoach, which was carried by the participants in their hands. It contained a collection of popular music, generally appreciated by students, covering a wide range of music tempos. The SportsCoach monitored heart rate, music tempo and running tempo throughout the experiment.

The participants did six separate sessions of 10 min each, spread over 2–3 days. Before each session they were given the opportunity for a 2-min warming-up, after which the 10-min actual monitoring session started. After a cooling-down period, they were asked to fill out a series of questionnaires (described above) on a laptop. Before the first session they were told that in all six sessions their task would be the same: to keep running such that their heart rate would remain in the range between 70 and 80% of their maximum heart rate, as calculated from their age by the formula HR_{max}=220-age. For each individual participant it was pointed out to which absolute heart rate values the range corresponded. The six sessions originate from the 3×2 experimental design, and were described as follows:

1. *No coaching & asynchronous music.* This condition is basically the same as regular running with music. The runners would have no indication at all what their heart rate was (except possibly introspection) and were advised to simply try and make the best of it.
2. *No coaching & synchronous music.* Similar as number 1, but now the SportsCoach was set to "Follow" mode, so that the music tempo followed the running tempo.
3. *Explicit feedback & asynchronous music.* Music tempo did not adapt, and the participants were given an additional Polar chest belt and wrist watch displaying their heart rate. The values between which their heart rate should stay were again pointed out to them.
4. *Explicit feedback & synchronous music.* Similar as number 3, but now the SportsCoach was set to "Follow" mode, so that the music tempo followed the running tempo.

5. *Implicit coaching & asynchronous music.* When the heart rate was above (below) the given optimal range, the music was played with a proportionally higher (lower) pitch. Thus the pitch of the music was varied, but the tempo remained at its original value. The participants were explained that in order to keep their heart rates in the optimal range, they only needed to increase their running tempo if the pitch increased, and vice versa.
6. *Implicit feedback & synchronous music.* This is the full SportsCoach version. It is the same as 5, except that now the music tempo changes in stead of the pitch. Accordingly, the participants were instructed that in order to keep their heart rates in the optimal ranges, they should keep running in sync with the music.

Half of the participants started with session 1 and ended with session 2, and the other half vice versa. The other 4 sessions were presented in between in a way counterbalanced over participants, in order to eliminate order effects.

8.3.4 Results and Discussion

A repeated measures ANOVA on the Borg scale data did not reveal a main effect for music mode, nor for coaching type. Neither did we find an interaction effect. Apparently the perceived exertion was not influenced by any of the conditions. This does not necessarily contradict the findings of Boutcher and Trenske, 1990 that music is beneficial to lower perceived exertion in low and moderate workouts, since we involved music in all our conditions.

The IMI results consist of 6 separate subscales (interest/enjoyment, perceived competence, effort/importance, pressure/tension, perceived choice and value/usefulness), which were analyzed in 6 independent repeated measures ANOVAs. For two of these we found a main effect of music mode: for interest/enjoyment, which is considered the main intrinsic motivation subscale ($F(1,11) = 6.3$, $p < 0.05$, see Fig. 8.4), and for perceived competence ($F(1,11) = 11.4$, $p < 0.05$, see Fig. 8.5). In both cases, synchronous music yielded higher scores: participants felt that synchronous music elicited more interest and enjoyment as well as enhanced their competence. These findings are in line with the conclusion of Anshel and Marisi, 1978 who emphasize the beneficial effects of music that is synchronous to sports movements. No main effects were found for coaching type in any of these ANOVAs, nor any interaction effects. Apparently the impact of coaching type is much less.

The Flow (FSS2) questionnaire consists of 9 different subscales, which probe different elements that can contribute to the experience of flow: challenge-skill balance, merging of actions and awareness, clear goals, unambiguous feedback, concentration on the task at hand, sense of control, loss of self-consciousness, transformation of time, and autotelic (intrinsically rewarding) experience. We again analyzed them in 9 separate repeated measures ANOVAs. The only main effect for coaching type was found for the unambiguous feedback subscale ($F(2,11) = 12.9$, $p < 0.05$). In

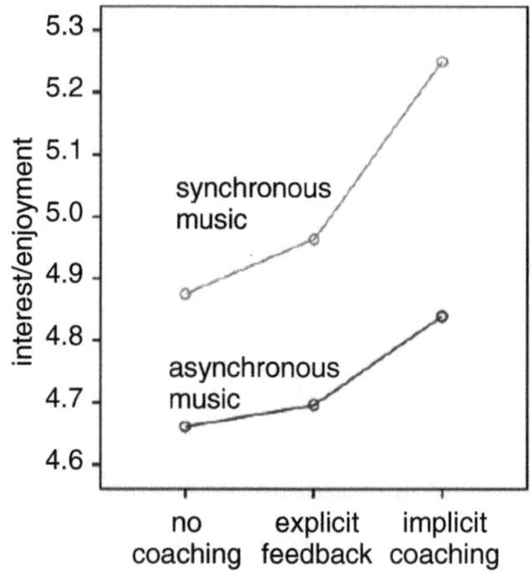

Fig. 8.4 Interest/enjoyment (IMI) as a function of coaching type and music mode

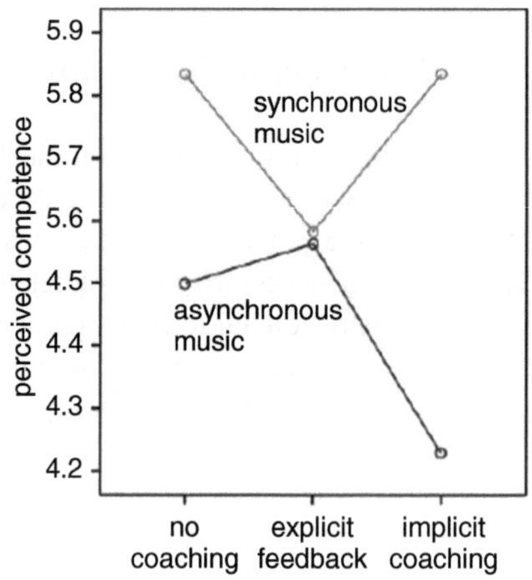

Fig. 8.5 Perceived competence (IMI) as a function of coaching type and music mode

Fig. 8.6, it can be seen that this effect is mainly due to the enhanced unambiguous feedback which is experienced in the explicit feedback conditions. And indeed, it makes sense that runners experience the highest amount of unambiguous feedback in the explicit feedback conditions. Furthermore, we found a main effect of music mode for four of the subscales: merging of action and awareness ($F(1,11) = 7.1$, $p < 0.05$), clear goals ($F(1,11) = 5.9$, $p < 0.05$), unambiguous feedback ($F(1,11) =$ 42,4, $p < 0.01$) and autotelic experience ($F(1,11) = 9.6$, $p < 0.05$). For all of these effects, we found that synchronous music enhanced the flow qualities, as can be seen from Figs. 8.6, 8.7, 8.8, and 8.9. As with the motivational data, these results on flow indicate the pleasant running experience elicited by synchronized music, demonstrating its ability to enhance the state of awareness of runners as well as the level of intrinsic reward experienced. Only for one subscale, clear goals, we found a significant interaction between music mode and coaching type (see Fig. 8.8). Apparently,

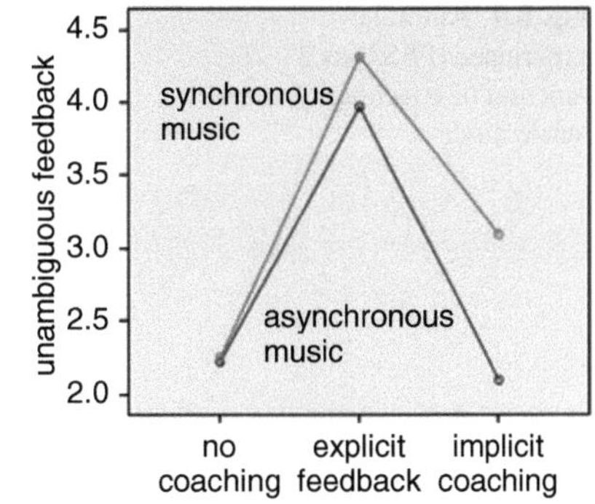

Fig. 8.6 Unambiguous feedback (FSS2) as a function of coaching type and music mode

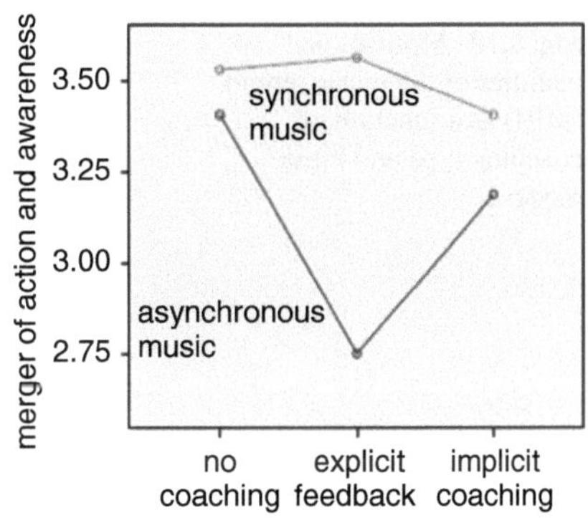

Fig. 8.7 Merger of action and awareness (FSS2) as a function of coaching type and music mode

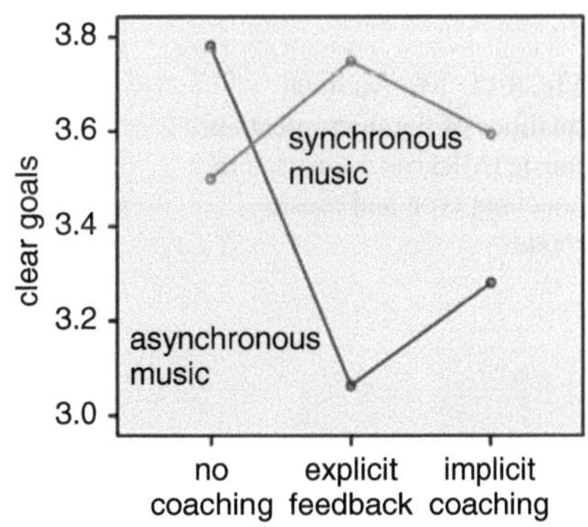

Fig. 8.8 Clear goals (FSS2) as a function of coaching type and music mode

the condition without feedback and with asynchronous music fosters the clearest goals in the participants, maybe because it is simply the familiar "running just like we always do".

The experienced motivational qualities of music (MRI) were scaled on 6 subscales (rhythm, style, melody, tempo, sound of the instruments, beat), for each of which we did a separate repeated measures ANOVA. The only main effect found

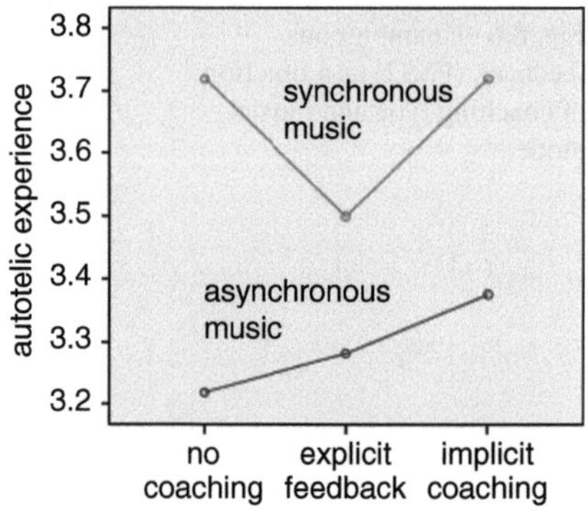

Fig. 8.9 Autotelic experience (FSS2) as a function of coaching type and music mode

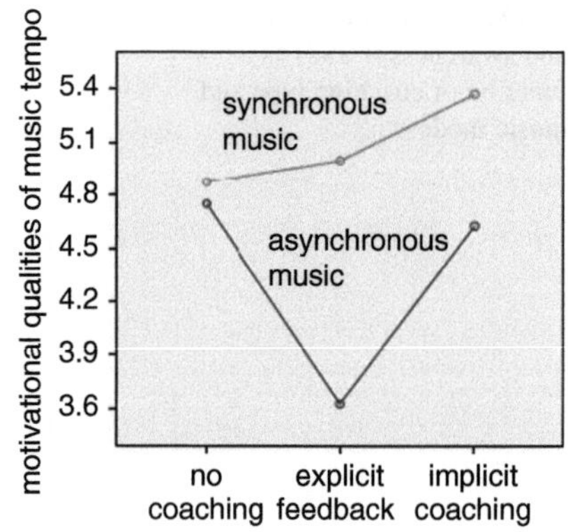

Fig. 8.10 Motivational qualities of the music tempo (MRI) as a function of coaching type and music mode

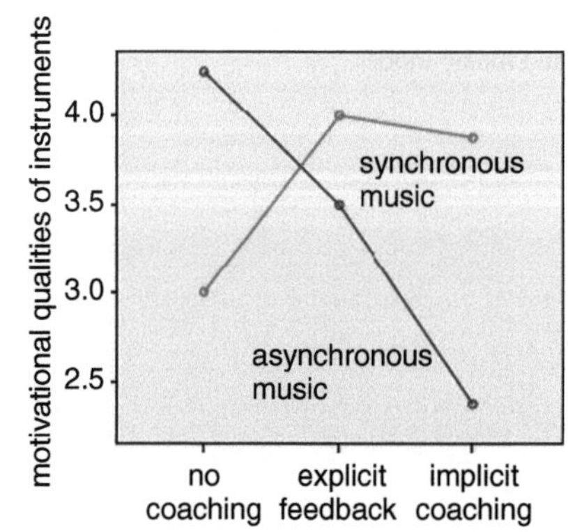

Fig. 8.11 Motivational qualities of the instruments in music (MRI) as a function of coaching type and music mode

was for the motivational qualities of the tempo of the music, which was significantly affected by music mode ($F(1,11) = 9.6$, $p < 0.05$, see Fig. 8.10). Again we find that, independent of coaching type, synchronous music raises the motivational qualities of the tempo of the music, which is in line with the conclusions of Anshel and Marisi (1978). This conclusion is strengthened by the fact that the effect only shows up for the tempo scale, and not for the other sub-scales. The only other significant effect that was found was an interaction effect between music mode and coaching type on the motivational qualities of the instruments in the music ($F(1,11) = 4.8$, $p < 0.05$, see Fig. 8.11). The interpretation of this interaction is not straightforward, however. Possibly, the effect is connected to the clarity of goals (Fig. 8.8), where we also find a high score for the asynchronous no-coaching condition ("plain normal running"), in that the clarity of goals leaves space to become aware of other aspects of the music than just tempo. That would still not explain, however, why this effect did not show up for the other subscales tested.

8.3.5 Conclusion

This experiment underlined the influence synchronous music has on the experience of runners: it motivates them, especially through its tempo, and raises their perceived competence. Furthermore, it enhances several aspects that are indicative of the experience of flow. The effect of coaching type was less pronounced, and especially the anticipated positive influence of implicit coaching was not supported by the results. Nevertheless, it appeared that our SportsCoach concept performed quite well on almost all aspects tested: Although its implicit type of coaching was not always unambiguous, it did score high on all other motivational and flow aspects.

8.4 Overall Discussion and Summary

Both experiments underline the potency of the SportCoach concept: adapting your running tempo to the music tempo is feasible, although possibly only within small ranges, and coaching through tempo changes is generally experienced in a positive way as well. Moreover, the synchronicity of music and running tempo adds to motivation and flow.

That being said, we also have to conclude that neither of the experiments supports that what we called implicit coaching was truly natural or effortlessly interpreted. In the first experiment, we had to conclude that adapting one's running tempo to a changing music tempo is not easy in many cases. In the second experiment, although the SportsCoach prototype scored relatively positive in many instances, we hardly ever found a significant main effect of coaching type, which could have indicated the benefits of implicit coaching per se.

In both experiments, the contact of the participants with the various conditions has been relatively short in comparison with the time runners usually spend running. And although the results obtained in these short encounters in the context of running are promising, it is interesting to explore the consequences of a longer exposure to "implicit" coaching through music tempo adaptation: Possibly, even the more experienced runners will learn to adapt their moves to slower and slower music tempos, just like dancers can. Possibly, the motivational advantages of the synchronous music will wine once the novelty of this feature has worn off, although on the other hand the implicit type of coaching might get more automated and internalized by the runners on the longer run, and therefore more truly implicit...

References

Anshel, M.H., Marisi, D.Q.: Effect of music and rhythm on physical performance. Res. Q. **49**(2), 109–113 (1978)

Boutcher, S.H., Trenske, M.: The effects of sensory deprivation and music on perceived exertion and affect during exercise. J. Sport Exerc. Psychol. **12**(12), 167–176 (1990)

Chen, M.J., Fan, X., Moe, S.T.: Criterion-related validity of the Borg ratings of perceived exertion scale in healthy individuals: a meta-analysis. J. Sports Sci. **20**(11), 873–988 (2002)

Csikzentmihalyi, M.: Flow: The Psychology of Optimal Experience. Harper Collins, New York, NY (1990)

Deci, E.L., Ryan, R.M.: Intrinsic Motivation and Self-Determination in Human Behavior. Plenum, New York, NY (1985)

IJsselsteijn, W.A., de Kort, Y.A.W., Westerink, J., de Jager, M.,, Bonants, R.: Virtual fitness: stimulating exercise behaviour through media technology. Presence: Teleoperators Virtual Environ. **15**(6), 688–698 (2006)

Intrinsic Motivation Inventory: http://www.psych.rochester.edu/ SDT/measures/word/IMIfull.doc. Accessed 20 Oct 2006

Jackson, S.A., Eklund, R.C.: Assessing flow in physical activity: the flow state scale 2 and dispositional flow scale 2. J. Sport Exerc. Psychol. **24**, 133–150 (2002)

Karageorghis, C.I., Terry, P.C.: The psychophysical effects of music in sport and exercise: a review. J. Sport Behav. **20**, 54–68 (1997)

Karageorghis, C.I., Terry, P.C.,, Lane, A.M.: Development and initial validation of an instrument to assess the motivational qualities of music in exercise and sport. J. Sports Sci. **17**(9), 713–724 (1999)

Large, E.W.: On synchronizing movements to music. Hum. Mov. Sci. **19**, 527–566 (2000)

Oliver, N., Kreger-Stickles, L.: PAPA: Physiology and purpose-aware automatic playlist generation. Proceedings of ISMIR 2006, 7th International Conference on Music Information Retrieval, Victoria, Canada, 8–12 Oct 2006

Tenenbaum, G., Lidor, R., Lavyan, N., Morrow, K., Tonnel, S., Gershgoren, A., Johnson, M.: The effect of music type on running perseverance and coping with effort sensations. Psychol. Sport Exerc. **5**(20), 89–109 (2004)

Westerink, J., De Jager, M., Van Herk, J., De Kort, Y., Ijsselsteijn, W.: The influence of virtual coaching on social presence and motivation in home fitnessing. Psychol. Health **19**, 181–182 (2004)

Wijnalda, G., Pauws, S., Vignoli, F., Stuckenschmidt, H.: A personalized music system for motivation in sport performance. IEEE Pervasive Comput., Spec. Issue: Pervasive Comput. Sports **4**, 26–32 (2005)

Chapter 9
Sleep in Context

Henriette van Vugt

Abstract Usually, the bed is where the day ends and a new day begins. During sleep, people are mostly unaware of the things that happen in the environment, and therefore psychologically, sleep separates one day from the next. For many, an "ideal" night of sleep consists of quickly falling asleep, sleeping through the night, and waking up refreshed and ready to face the day (e.g., Taylor et al., 2008). However, some nights are not that ideal. Not only people with clinical conditions or sleep disorders, but also healthy people might sometimes have difficulties falling asleep and staying asleep, and wake up too early or unrefreshed (e.g., NSF, 2008; Cuartero and Estivill, 2007; Bixler, 2009). Many people without chronic sleep complaints also sometimes feel the need to be assured that the upcoming night will be a refreshing one, without troubles. Therefore, this chapter focuses on the sleep of healthy individuals.

9.1 Introduction

In this chapter, I will focus individual sleep-related experiences of people in their daily lives, and when appropriate, I will link these to scientific evidence. Several important topics will be addressed: why sleep matters to people (Section 9.2), how the mind and the body affect sleep (Section 9.3), and how the environment affects sleep (Section 9.4). Further, I will provide an overview of the pros and cons of existing measurement techniques for applications and evaluations in home contexts in which healthy individuals relax and sleep (Section 9.5).

Even though many people have problems falling or staying asleep, at least once in a while, most of them are reluctant to use sleeping pills, because of possible dependency of the body on the drug and other negative side-effects. Non-pharmacological treatments might be just as effective in the treatment of sleep difficulties and do not have the undesirable side-effects that hypnotics have. This chapter can be a useful starting point for designing non-pharmacological treatments for promoting sleep that healthy individuals will appreciate.

H. van Vugt (✉)
Brain, Body and Behavior Group, Philips Research, Eindhoven, The Netherlands
e-mail: henriette.van.vugt@philips.com

J. Westerink et al. (eds.), *Sensing Emotions*, Philips Research Book Series 12,
DOI 10.1007/978-90-481-3258-4_9, © Springer Science+Business Media B.V. 2011

9.2 Why Sleep Matters to People

9.2.1 Sleep Is Perceived as Important

People sleep about one third of their lives, and we sleep every day. Therefore, it is not surprising that people buy good mattresses, pillows, and other necessities to increase the chance of a good night's sleep (e.g., Cuartero and Estivill, 2007). People also discuss their sleep, for example on discussion forums such as Sleepnet[1] and Topix.[2] They talk about how well or how bad they slept last night. They talk about their dreams and nightmares. These are indications that people perceive sleep as a relevant component in their lives.

Many people like sleep. For example, the "Sleep in America Poll" reported an average sleep time of 45 min longer on non-workdays than workdays (NSF, 2008). For many people it is a treat to sleep in on weekends. People like the coziness of their bedrooms, the warmth of their beds. The bedroom can give people a feeling of safety and peace – although especially women can experience difficulties sleeping due to an "unsafe" feeling such as the worry about potential burglars (e.g., Van Vugt et al., 2009). Many people regard sleep as a positive experience, more than just a necessity. However, people also complain about sleep (as reported in e.g., Epstein and Mardon, 2006; Du et al., 2008), which can be seen as another indication that healthy people perceive sleep as important. People may complain about having problems with falling asleep at night, for example because of cold feet, light, or because they have difficulties to "switch off" their minds. Other people complain about waking up during the night. Many people like to sleep with an open window for fresh air, but then also outside noises may enter the bedroom which may wake people up. And finally in the morning, people may not wake up feeling refreshed, and others may wake up too early not being able to sleep again. Couples usually want to sleep in the same bed, which not only has positive but also negative consequences (e.g., Strawbridge et al., 2004). Partners may disturb each other because of different sleeping schedules. One person might work in night shifts while the other does not. One person might prefer going to bed early, while the other prefers going to bed later. Or one person might like to read or watch television in bed while the partner would like to sleep. Other sleep disturbances can be caused by partner's movements and snoring (e.g., Pankhurst and Horne, 1994; McArdle et al., 2001; Du et al., 2008).

People may also worry about sleep. For example, people worry whether they get sufficient sleep, especially if they sleep less than the "magic number" of 8 h per night (e.g., Epstein and Mardon, 2006, p. 27). They worry if they do not sleep before a certain time and if they have to get up early the next day (e.g., Du et al., 2008). And unfortunately, the more people worry about their sleep, the worse they often actually sleep. People might use sleeping pills, but then they might worry

[1] http://www.sleepnet.com/

[2] http://www.topix.com/forum/health/sleep

about health consequences and whether they might become dependent on the drug (e.g., NSF, 2008).Worries and intrusive thoughts are especially salient in insomnia patients (e.g., Wicklow and Espie, 2000; Harvey et al., 2005).

To conclude, there are several indications that people find sleep important. People hope that each night brings them a good night's sleep but environmental, psychological and even physiological disturbances may keep them from doing so. And rightly so, as there is abundant scientific evidence that sleep is important to people's health and wellbeing. For example, after a bad night's sleep, people may have difficulties to concentrate, perform well and learn (e.g., Dinges and Kribbs, 1991; Dinges et al., 1997; Walker, 2008). Poor sleep may also impair the decision making process (e.g., Venkatraman et al., 2007), and negatively affect the immune system (e.g., Opp, 2009).

9.2.2 A Balanced Lifestyle

Even though people know the importance of good sleep, they do not always act accordingly. For example, people sleep late and get up early because of social events and commitments such as work. People have to fit all daytime activities and time for sleep in the 24 h of a day (e.g., Broman et al., 1996). People say they want and need more sleep, and this has also been argued in literature (e.g., Broman et al., 1996; Spiegel et al., 1999; Dement, 2005; Bixler, 2009). On the other hand, one study showed that only few people opted for more sleep when given attractive waking alternatives (Anderson and Horne, 2008). This raises the question whether it actually is more sleep that people want. Another study comparing the effects of different types of free-time activity (work, quiet leisure activity, active leisure activity) on sleep, recovery and well-being (Tucker et al., 2008), showed interesting results. Evening activities involving relatively low mental effort were associated with improved subsequent self-reported sleep. Being satisfied with one's evening activities was also related to better self-reported sleep. Rest, recuperation and satisfaction were rated lowest in the work condition. Thus, when people say they need more sleep, they might actually mean they need more "time out" from their stressful (work) life that is full of demanding activities (Anderson and Horne, 2008). Whereas a *balanced lifestyle* often refers to the balance between work and leisure, a truly balanced lifestyle is not just about activities, but also about sleep and rest.

That people encounter difficulties having a balanced and healthy lifestyle can be caused by a range of factors. The way things are organized in society affects people's lifestyle, and this societal organization might not be ideal for each individual. For example, it is shown that the peak of alertness in morning type people occurs late in the morning, while in an evening type it occurs in the late afternoon (Natale and Cicogna, 1996). An evening type might feel and function better during the day when sleeping until later in the morning – and work until later in the evening – but early office hours may prevent him or her from doing so. In addition, certain jobs require people to work many hours without resting, or work during the night.

Other social factors also play a role. For example, people might like to take a short nap after lunch, but this is not always socially accepted. A Dutch website[3] reports that in Western-Europe, sleeping during the day is associated with laziness, and that a daytime nap is considered to be only for elderly and sick people. It even reports that "naps clash with our culture." Thus, in many countries and cultures, daytime sleepiness is considered a weakness and not something one should give in to by daytime napping. However, the discrepancy between the optimal biological sleep-wake schedule and the actual sleep-wake schedule – whatever the reason of occurrence – might negatively affect people's mood and performance (e.g., Dinges et al., 1997; Åkerstedt and Wright, 2009).

Further, to obtain a balanced and healthy lifestyle, people need to listen to their bodily signals of sleepiness. However, their minds often have a different plan. Even though people may be tired, some are tempted to think they can cope with sleep debt and function normally whereas their bodies actually can't. For example, people go to bed late for different reasons. Before the day of an exam people might study until late, even though they feel tired. Contrary to people's expectations, they may pass the exam yet lose much of what they "crammed into their brains" (e.g., Stickgold, 2005). Second, in the morning, many people need an alarm clock to wake them up, often as a consequence of staying up late. This can be seen as a sign that people have not yet slept enough – otherwise they would have woken up naturally. As indicated before, sleep dept can have many negative consequences. Third, drivers do not always notice their sleepiness and carry on driving (e.g., George, 2003). Car accidents tend to peak in the early morning when the body is inclined to rest and subjective alertness and performance are low (e.g., Walsh et al., 2005). Noteworthy is that in New Jersey in the United States drowsy driving is now treated as a criminal offense (e.g., Avara, 2008; Walsh et al., 2005). In general, people's mind can overrule bodily signals of sleepiness, and the resulting behavior may have unwanted negative effects on the longer term.

In sum, whereas people might strive for a balanced lifestyle, it is difficult to have one in our busy, 24 h society.

9.2.3 Mood and Sleep

The expression "to get up on the wrong side of the bed" indicates that sleep is related to one's mood. It is well-known that patient with mood disorders often have difficulties sleeping (e.g., NSF, 2009). But also in healthy people, mood is related to sleep. Indeed, studies have shown that sleep may negatively affect mood when sleep is restricted or people are sleep deprived (e.g., Franzen et al., 2008; Haack and Mullington, 2005). Further, even when the sleep of participants is not restricted, sleep affects day-to-day fluctuations in mood, despite all kinds of events and impressions that also influence mood during the day (e.g., Vossen et al., 2009). Also after

[3] http://www.lerenslapen.nl/page/1515/overdag-slapen-middagdutje.html

a bad night's sleep, people might not feel refreshed in the morning, affecting mood (e.g., Stone et al., 2008). Not only objective parameters of sleep such as total sleep time, but also people's *subjective perceptions* of sleep and sleepiness affect mood (e.g., Boivin et al., 1997). One study showed that subjective ratings of cheerfulness and happiness gradually decline during wakefulness and the poorest mood ratings occur in the middle of the biological night (Boivin et al., 1997). Another study in which the sleep of participants was not restricted, showed that mood positively correlates with rates of subjectively sleep quantity and quality, but less so with objective sleep parameters such as total sleep time and sleep onset latency (Vossen et al., 2009). Thus, subjective perceptions of sleep seem more important for mood than objective parameters.

The process of waking up may also affect mood. Many people dislike waking up by a beeping alarm clock, and prefer waking up by their own "biological clock". Some people like to wake up at the end of a dream – at least if it is not a nightmare – to be able to remember it (e.g., Du et al., 2008). However, whereas some people remember a dream almost every day, others almost never remember a dream (e.g., Cohen, 1974; Fitch and Armitage, 1989). Strategies to ease the recall of dreams can be found in scientific literature (e.g., Yu, 2006), and also on the web, for example on the site "Ten tips for dream recall".[4] Scientific research shows a relation between happy dreams and happy waking emotions (Cartwright, 2005; Gilchrist et al., 2007). Thus, the recall of a happy dream makes the start of the day more pleasant.

Many people take countermeasures to prevent sleepiness and to increase daytime alertness levels – hereby also affecting mood. Common countermeasures to fight daytime fatigue are coffee and other caffeine-rich drinks. Naps may also help fight sleepiness (e.g., NSF, 2008; Horne et al., 2008; NSF, 2008), although some people say that napping for too long causes them to feel even more tired (e.g., Du et al., 2008). Whereas many people like napping or having a siesta, few people take the time to do so, increasing the popularity of caffeine-rich drinks (e.g., Cuartero and Estivill, 2007, p. 45). People may also perform exercise when feelings sleepy during the day, or just accept the sleepiness and keep going on – perhaps going to bed early that night (NSF, 2008). Some people like to sleep longer on weekends (NSF, 2008), however, this may destabilize the sleep-wake rhythm and people may find it hard to get up earlier again on Monday morning (e.g., Taylor et al., 2008).

In conclusion, considering that they experience negative consequences of a bad night's sleep directly themselves, and are willing to take all types of countermeasures, it is not surprising that people find sleep important. In the following section, I will describe the issues surrounding the body and the mind people perceive as sleep thieves.

---

[4]http://www.selfgrowth.com/articles/Ten_Tips_for_Dream_Recall.html

9.3 How the Mind and the Body Affect Sleep

9.3.1 The Mind

A relaxed mind helps people fall asleep (cf. Tucker et al., 2008). However, for people with a busy and stressful life it is often difficult to find time to relax. A bedtime routine is a strategy to wind down and relax both the mind and the body before going to sleep, which may subsequently affect sleep and sleep onset. However, people do not always succeed to follow such a routine. As a result, many healthy people experience trouble falling asleep when they are stressed and their minds are full of thoughts, ideas, and worries (e.g., Bonnet, 2000; Åkerstedt, 2006; Du et al., 2008; Bixler, 2009). A busy mind can be caused by stressful events at work or at home, and worries about children and family matters (e.g., Urponen et al., 1988). And even nice thoughts or ideas that occupy the mind, unrelated to stress or worries, can keep people from falling asleep. For others, an untidy house can cause a feeling of restlessness, which only seems to disappear when the house or room is tidy, and everything is in place. The feeling of safety also contributes to a relaxed mind. Worries about the house and burglars can keep people awake (Du et al., 2008).

Depending on the cause of the restless mind, people use many different strategies to calm their minds before going to sleep (Urponen et al., 1988; Epstein and Mardon, 2006; Du et al., 2008). Reading is a popular bedtime activity among healthy individuals, as well as insomnia patients (Morin et al., 2006b). Other bedtime routines to calm or distract the mind may consist of drinking tea, watching television, taking a bath, or whatever helps people to unwind and calm their minds, such as talking with someone and preparing for the next day (e.g., cloths and bag). A bedtime routine can also include locking doors and windows in the house. Sometimes, it is the feeling of another person "being there" that is comforting and increases the feeling of safety. After a quarrel or argument, people may need to finish a fight for peace of mind. A strategy that many people use while lying in bed, is writing thoughts on a piece of paper to empty their minds and not forget about them the next day. Furthermore, music can reduce anxiety and counteract mental arousal before sleep (e.g., Evans, 2002). Music is indeed used often in the home context for relaxation purposes (e.g., Morin et al., 2006b; Urponen et al., 1988). Slow rhythm music, without a heavy beat, is experienced as relaxing by many people, but the effect is strongly dependent on personal preferences (De Niet et al., 2009).

When people are "forced" to wake up especially when they just have fallen asleep, for example because of people talking loudly or by a car alarm, they can get irritated. The irritated mind may prevent them from falling asleep again, even when the noise has already faded away.

When people are awake in the middle of the night, people may get up and eat or drink something, go to the toilet, read something, watch TV, or perform relaxation exercises (Epstein and Mardon, 2006; Du et al., 2008). When lying in bed without sleeping quickly – within about 20 min –, it is recommended to leave the bed.

This helps to associate the bed with sleep, not with negative associations of irritation, frustration and worries about sleep. This method originates from the 1970s and applies stimulus control theory to treat insomnia (Bootzin, 1972; Bootzin and Perlis, 1992; Manber and Harvey, 2005).

9.3.2 The Body

This section focuses on the many ways that the body may affect sleep.

The preparation for a good night sleep already starts at daytime. Most importantly, physical effort, if timed correctly, affects sleep. Historically, daytime exercise has been closely related to better sleep – more than any other daytime behavior (e.g., Youngstedt, 2005). Overall, research has repeatedly shown that exercise provides three critical benefits for sleep: people fall asleep faster, attain a higher percentage of deep sleep, and awaken less often during the night (Epstein and Mardon, 2006). For example, one study found correlations between sleep problems and decreased exercise, based on self-report of sleep disturbance and frequency of exercise (Bazargan, 1996). Physical activity may also hamper sleep, especially sleep onset, when performed too late at night. Indeed, people report that exercising too heavily or too late in the evening disturbs sleep (e.g., Urponen et al., 1988). Physical exercise has an arousing effect, and therefore does hamper the body in its preparation for sleep. Timing of physical exercise is therefore important if one wants to obtain beneficial effects. However, even if people would like to schedule sports earlier in the day, work and other obligations may prevent them from doing so and therefore sports are often scheduled in the evening hours (e.g., Bureau of Labor Statistics, 2008).

Sports may also negatively affect sleep if injuries and aches are involved (e.g., Jennum and Jensen, 2002; Fahlström et al., 2006; Gosselin et al., 2009). Pain does not only disturb sleep continuity and sleep quality, but poor sleep also further exacerbates pain (Smith and Haythornthwaite, 2004), which makes things worse.

Further, sometimes people cannot find the right position to fall asleep. People experience the effect of an inadequate body position on sleep while traveling and trying to sleep in a bus or airplane. An upright position may also antagonize sleep (e.g., Bonnet, 2000). People have less difficulties sleeping in a bed than in a chair, or worse, while standing. To promote sleep onset, a horizontal posture is best. Neck and chest should be below the heart, and legs above the heart (Cole, 2005).

Another sleep thieve strongly related to the body is muscle tension. The muscles in the body can be tense for different reasons such as stress or intensive computer use. Muscle relaxation is a technique to relieve stress, and subsequently helps people fall asleep faster (e.g., in insomnia patients, Morin et al., 1999). Some people like to give each other a massage before sleep (e.g., Du et al., 2008), which is one form of muscle relaxation. Another relaxation technique is called progressive muscle relaxation (tensing and relaxing different muscle groups in sequence). Slow and regular breathing, also called paced breathing, may also help to relax the body. One study showed that slow and regular breathing, guided by music, can lower blood pressure and heart rate (e.g., Grossman et al., 2001). Although paced breathing might be an effective method for winding down in the evening after stressful events, I am not aware of studies investigating people's personal experiences with and willingness to perform breathing exercises in their homes to relax and wind down. Other relaxation techniques to help people fall asleep faster are visualization exercises, mediation, yoga, biofeedback and autogenic training. However, not all people might benefit from relaxation techniques prior to sleep. If relaxation therapy is used in people that are already relaxed prior to sleep but cannot sleep, sleep issues might even become worse (Hauri et al., 1982).

Further, many people, mostly women, can have difficulties sleeping with a cold body, especially cold feet and hands (e.g., Krauchi et al., 1999, 2000). Different strategies are employed to tackle this sleep thieve. People might take a hot bath or shower prior to sleeping (e.g., Liao, 2002), put on bed socks, or, when sleeping in the same bed as their partner, use the warmth of the partner to warm up (e.g., Cole, 2005; Du et al., 2008). Research has shown that warming up cold feet or other body parts may indeed make people fall asleep faster (e.g., Liao, 2002; Sung and Tochihara, 2000; Van Someren et al., 2002; Raymann et al., 2007). Temperature change may also be accomplished by sports and physical activity. It could be that while exercising, people warm up, and once finished with exercising, people cool down, and that this process of cooling down prepares the body for sleep (e.g., Van Someren, 2000).

Last, people experience that alcoholic drinks, coffee, and eating too heavily or too late at night may disrupt their sleep (e.g., Du et al., 2008). While awake in the evening, the alcohol or food may seem attractive, but while lying in bed, people may feel the negative consequences of consuming too much late at night. Even though a "nightcap" is considered to improve relaxation on falling sleep, it is indeed a disturbing factor for sleep continuation (e.g., Upsonen, 1988; Bixler, 2009). Not all consumption is considered disruptive for sleep. In the evenings, some people like to drink warm milk or herbal teas, and avoid drinking coffee to facilitate sleep (e.g., Du et al., 2008), and insomnia patients also use herbal or dietary products to fall asleep (e.g., Morin et al., 2006a). Sleep hygiene education includes general guidelines about health practices (e.g., diet, education exercise, substance use), as well as environmental factors (e.g., light, noise, temperature) that may promote or interfere with sleep (e.g., Morin et al., 2006a), which will be discussed in the following section.

9.4 How the Environment Affects Sleep

When asked about the optimal environment to sleep, people often refer to the importance of darkness (e.g., Urponen et al., 1988; Bonnet, 2000; Van Vugt et al., 2009). For some, even a small red light of a television or mobile phone is disruptive for sleep. Even though people can use eye masks to ensure darkness, this is not an often used method. Indeed, even low light conditions inhibit endogenous melatonin production, which is an important sleep promoter (e.g., Czeisler et al., 2005).

Further, people also report the importance of tranquility for sleep (e.g., Urponen et al., 1988; Bonnet, 2000; Van Vugt et al., 2009). Noises of outside traffic, people talking in the street, a snoring partner, a crying or restless infant, a washing machine, a television, or air conditioning in hotels can all make it difficult to fall and stay sleep. Indeed, research has shown that even a mild sleep disruption that suppressed deep sleep, but did not reduce total sleep time, was sufficient to affect memory performance in healthy human subjects (Van der Werf, et al., 2009). Disturbances of light and noise can also be caused by a partner who sleeps in the same bedroom. While one may like to read, or watch television, the other may be annoyed by the light and noise (e.g., Du et al., 2008). Snoring may also severely disturb the sleep of the other person (e.g., Venn, 2007). Adults may use earplugs or close the windows of the bedroom for noise reduction. However, not all noise may disrupt sleep. As indicated in an earlier section of this chapter, music may facilitate sleep through mental relaxation. In addition, white noise and noise with a rhythmic sound may help masking other noises (e.g. Cole, 2005), which is a technique commonly used to soothe infants.[5] Interestingly, people can have cat naps during the day when it is light and noisy! In general, though, sensory withdrawal in the bedroom is recommended, which includes the absence of light and noise (including television), as well as comfort and stillness, and a tidy sleeping environment (e.g., Cole, 2005).

The climate in the bedroom is another environmental factor that affects people's bedroom experience. The preferred room temperature differs individually – some like their bedroom to be cold and like to sleep with their windows open, even in winter, while others like to sleep with the heating on (e.g., Du et al., 2008). Heating systems may dry the air in the bedroom, and a dry throat may make sleeping more difficult. Humidifiers may then be useful.

Last, smell affects people's bedroom experience. Smells may help or hinder people to relax and fall asleep. Normally, people cannot keep their noses from smelling and hence (bad) smells cannot be disregarded. Lavender or rose smells are associated with relaxation, and some people use essential oils with these particular smells in their bedrooms (e.g., Du et al., 2008). Smell has not extensively been studied in the context of sleep, but one study found that information processing of smells is present in sleep and that the emotional tone of dreams can be influenced depending on the type of smell (Schredl et al., 2009). The effect of smells on people's wake up experience has not been extensively studied.

Another factor with a clear link to body comfort and sleep quality is the bedding system. Because humans shift their body position between 40–60 times per night, they need a good mattress and pillow, as well as plenty of room to move (e.g., Cuartero and Estivill, 2007). People think that soft sheets and pillows may also increase the comfort of the body and subsequently contributes to a restful mind (e.g., Cuartero and Estivill, 2007). Indeed, bedding systems and the firmness of the mattresses affect back pain and sleep quality (e.g., Garfin and Pye, 1981; Jacobson et al., 2009). However, there are no clear guidelines on who should use what type

[5] http://www.babyslumber.com/white-noise

of bedding and mattress for maximal comfort and sleep quality, and reduction of sleep disturbances (Jacobson et al., 2009). Despite the lack of clear guidelines, the bedding and mattress industry is huge and its sleep claim largely depends on marketing.

Environmental factors are often related to the socio-economic circumstances people live in. Arber et al. (2009) analyzed a British nationally representative survey of over 8,000 men and women aged 19–74 to investigate in sleep problems. Other than generally believed, they showed that the gender difference in the amount of sleep problems (women report more sleep problems than men) was not due to differences in health, health worries, chronic illness, and depression, but to the more disadvantaged socio-economic circumstances of women. Low socio-economic status is associated with living in smaller, poorer quality housing, with fewer and more shared bedrooms and insubstantial walls, and with living in disadvantaged neighborhoods (Arber et al., 2009). This often goes hand in hand with higher noise levels and a climate that is not ideal for sleep.

9.5 Techniques for Measuring Sleep in People's Homes

This section describes the various measurement techniques for evaluating sleep of healthy people in home contexts. Sleep, sleep quality, sleepiness, and other sleep-related factors can be measured subjectively and objectively. The various measurement techniques can be used on their own or together to deepen understanding and for validation purposes, for example by gathering both quantitative and qualitative data on a certain topic.

Goals and questions should lead all studies. Thus, before using any measurement technique, either subjective or objective, the goal of the study should be clear. The design of the study should be thought through. However, there is often no need to articulate the goals to the people under investigation as long as the researcher knows the goals and questions the study should focus on. This is even true for field studies that have a more open character in which the researcher should carefully balance between being guided by goals and being open to modifying, sharpening, and refocusing the study after learning about the situation (e.g., Preece et al., 2002).

9.5.1 Objective Measurements

Typical objective sleep parameters are total sleep time (TST), number of awakenings, sleep-onset latency, wakefulness after sleep onset, early morning awakening, and sleep efficiency (which is the total sleep time divided by the time in bed), as well as sleep architecture (stages), REM sleep, and respiratory and movement parameters. Whereas the "golden standard" to measure such sleep parameters is polysomnography (PSG) using electrodes on the head, this is not a preferred method to study sleep in home contexts because of its obtrusiveness and relative difficulty to use. Recently, the Zeo company[6] has launched a simplified, PSG-based, device for measuring sleep in home by means of a wireless headband with integrated EEG sensors. Often used devices for home monitoring of sleep are wrist actigraphs, for example the actiwatch of Philips Respironics.[7] These are small computerized devices that record and store data generated by movements of the arm. Some actigraphs include a light sensor and a skin temperature sensor for additional information. Movements and activity during sleep can also be recorded by video-actigraphy (e.g., Liao and Yang, 2008). Actigraphs are considered reliable for sleep assessment (Tryon, 2004).

Another typical sleep parameter is daytime sleepiness. The Multiple Sleep Latency Test (MSLT) is the standard way to measure a person's level of daytime sleepiness. It can be used to see how quickly people fall asleep in quiet situations during the day. Additional objective sleep related measurements are "lights off, lights on", and alcohol and medication intake.

Different parameters are especially important for studying the human circadian clock. Important markers of the circadian phase are Dim Light Melatonin Onset (DLMO) and Core Body Temperature. DLMO can be determined by means of hormomal measurements in saliva, plasma, or urine (e.g., Pandi-Perumal et al., 2007). CBT can be determined by rectal measurements (as e.g., in Raymann et al., 2005) or a CBT pill that can be swallowed.

9.5.2 Subjective Measurements

Three main categories of subjective measurement techniques for sleep can be distinguished: (1) questionnaires, (2) interviews, and (3) observations.

9.5.2.1 Questionnaires

Questionnaires are a well-established technique for collecting demographic data and people's subjective views. Questionnaires can be printed on paper and used in a laboratory setting or send to people by mail. They can also be administered online to more easily reach a larger number of people. It is important that questionnaires reach a representative sample of participants and that a reasonable response rate is ensured. Thus, researchers should specify the target audience and think about how and when these people can be reached, and about how to encourage good response (e.g., Preece et al., 2002).

Different standardized questionnaires are available to measure introspective sleepiness and other sleep-related factors (e.g., Bae and Golish, 2006; Lomeli et al., 2008). For example, the Stanford Sleepiness Scale (SSS) has been in the running

[6] http://www.myzeo.com/

[7] http://www.actiwatch.respironics.com/

for many years to measure introspective sleepiness (Hoddes et al., 1973). People choose between seven statements to describe their self-assessed current sleepiness state, that vary between "Feeling active, vital, alert, or wide awake" to "No longer fighting sleep, sleep onset soon; having dream-like thoughts". Further, the Epworth Sleepiness Scale (ESS) measures the self-reported expectation of dozing off or falling asleep in a variety of situations, such as sitting and reading, watching television, sitting quietly after a lunch without alcohol; in a car, while stopped for a few minutes in traffic (Johns, 1991). The Pittsburgh Sleep Quality Index (PSQI) was developed to measure sleep quality during the previous month and to discriminate between good and poor sleepers (Buysse et al., 1989). Sleep quality is a complex phenomenon that the PSQI measures by several dimensions, including subjective sleep quality, sleep latency, sleep duration, Habitual sleep efficiency, sleep disturbances, use of sleep medications, and daytime dysfunction. People's diurnal preference can also be measured, for example by the Munich ChronoType Questionnaire (MCTQ; Roenneberg et al., 2007) or the Morning-Eveningness Questionnaire (Horne and Ostberg, 1976; Roenneberg et al., 2003). Last, social factors can be determined through the Social Rhythmic Questionnaire (Monk et al., 1990).

To my knowledge, no standardized questionnaire measures the sleep *experience* of healthy people in their own homes. Such a questionnaire may cover aspects such as those described in "the people's perspective" sections of this chapter, in order to understand (a) the importance of sleep in people's daily lives, (b) the aspects related to the body that affect people's sleep, (c) mental aspects that affect people's sleep, and (d) environmental aspects that affect people's sleep.

9.5.2.2 Interviews

Despite the different goals that interviews might have, all interview situations have in common an interviewer and a respondent engaging in social exchange. Already in 1924, Bingham and Moore referred to the interview as a "conversation with a purpose" (Bingham and Moore, 1924). Different categories of interviews are named according to how much control the interviewer imposes on the conversation: unstructured, structured, and semi-structured. A last category is group interviews where the interviewer facilitates discussion among a small group of people (Fontana and Frey, 1994, see also Preece et al., 2002). Focus groups are one type of group interview in which a small, representative group of people with a similar background are gathered for a group discussion on preset topics, guided by a facilitator. The pros and cons of, and issues involved in each type of interview are explained in an easy manner in Preece et al. (2002).

In many interviews, interviewer and interviewee are in the same location, for example in people's homes, in a laboratory, or on the street. Interviews can also be conducted via the telephone, videoconferencing or online, if it is unfeasible or unpractical to meet. An advantage of face-to-face interviews is that it allows for interpreting body language in the context of the conversation. An advantage of online interviews is that (sensitive) questions can be answered anonymously.

In the context of sleep, interviews are often conducted in sleep clinics with patients with sleep disorders for diagnostic purposes (e.g., Bae and Golish, 2006). As a noticeable exception, Arber et al. (2009) report about a large dataset of interviews with people without sleep disorders, conducted in their homes. Their objective was to assess whether social factors mediate gender differences in sleep quality. They found that disadvantaged socio-economic characteristics are strongly related to sleep problems.

Almost every year, the American National Sleep Foundation conducts survey research on about 1,000 telephone interviews. The objectives of the 2008 research were, among others, to get insight into the sleep habits of working Americans, the relation between sleep habits and work performance, the number of working Americans that experience sleep problems (NSF, 2008).

9.5.2.3 Observations

Data from focus groups, interviews and questionnaires provided understandings of what sleep means in people's lives, a disadvantage is that these techniques are potentially subject to recall bias, as the participant's input is "distanced by time from the temporal, spatial and relational realities of sleep" (Hislop et al., 2005). There are several observational techniques that can tackle one or more of these disadvantages.

Watching and listening to people can tell much about what they do, in the context in which they do it, how well technology supports them, and what other support might be needed (e.g., Preece et al., 2002). Depending on the goals of the observation and the questions addressed, it might be useful to observe people in their own home environment, in public spaces, in a laboratory environment, etcetera. For example, a researcher aiming to better understand the bedtime routines of children might decide to do a study at people's homes to be able to observe the family during the bedtime routine, involving the living room, child's bedroom, and other relevant spaces. Conducting studies in home environment on the topic of sleep requires special attention to the participants' privacy. A researcher aiming to test the usability of the Philips' Wake-up lamp might decide to observe people while using it in a laboratory environment. The level of participation of the observing researcher can vary, from on the one extreme being complete participants themselves, to the other extreme observing from the outside without participation (e.g., Robson, 1993).

A researcher can use a variety of observation techniques, such as think-aloud protocols that help participants verbalize their thoughts, actions, and experiences (Erickson and Simon, 1985) and frameworks that helps researchers to keep their goals and questions in mind (e.g., Preece et al., 2002). Further, in a "context-mapping" study (Sleeswijk-Visser et al., 2005), designers and researchers aim to gain deeper understanding of the needs and dreams of prospective users of new products by involving participants intensively in creating an understanding of the contexts of product use. In the context-mapping study we performed (Du et al., 2008; Van Vugt et al., 2009), interviews, observations, and other research tools such as diaries and creative exercises were combined for a more comprehensive understanding of the context in which participants reside and their behavior and experiences. During observation, both note taking and video- and audio taping are recommended for later data-analyses. The pros and cons of note-, video-, and audio-based techniques are summarized in Preece et al., 2002, p. 276).

Last, sleep diaries are widely used instruments to observe sleep problems and sleep experiences, measuring factors such as subjective sleep timing and subjective sleep quality. The value of diaries is their potential to record events, over time, as close as possible to when they occur (Elliot, 1997). Two well-known and widely used sleep diaries are *The Pittsburgh Sleep Diary* (Monk et al., 1994) and the sleep diary by Lichstein et al. (1999) that are filled out by participants on paper, typically without the researcher being present. Hislop et al. (2005) have used audio sleep diaries in researching sleep. They argued that the technique gives "insights into sleep experiences on a nightly basis as close to the event as possible and independent of researcher involvement, thus ensuring relatively untainted individual interpretations of the experience of sleep". Paper and audio sleep diaries can complement each other in sleep research (see Hislop et al., 2005).

References

Åkerstedt, T.: Psychological stress and sleep. Scand. J. Work Environ. Health **32**(6), 493–501 (2006)

Åkerstedt, T., Wright, K.P., Jr.: Sleep loss and fatigue in shift work and shift work disorder. Sleep Med. Clin. **4**(2), 257–271 (2009)

Anderson, C., Horne, J.A.: Do we really want more sleep? A population-based study evaluating the strength of desire for more sleep. Sleep Med. **9**, 184–187 (2008)

Arber, S., Bote, M., Meadows, R.: Gender and socio-economic patterning of self-reported sleep problems in Britain. Soc. Sci. Med. **68**, 281–289 (2009)

Avara, E. The correlation between sleep apnea & automobile crashes. Business people. URL: http://findarticles.com/p/articles/mi_qa5341/is_200809/ai_n29494028/ (Sept 2008)

Bae, C.J., Golish, J.A.: The sleep interview and sleep questionnaires. In: Lee-Chiong, T. (ed.) Sleep: A Comprehensive Handbook, pp. 967–971. Wiley, Hoboken, NJ (2006)

Bazargan, M.: Self-reported sleep disturbance among African-American elderly: the effects of depression, health status, exercise, and social support. Int. J. Aging Hum. Dev. **42**(2), 143–160 (1996)

Bingham, W.V., Moore, B.V.: How to Interview. Harper & Brothers, New York, NY (1924)

Bixler, E.: Sleep and society: an epidemiological perspective. Sleep Med. **10**, 1–4 (2009)

Boivin, D.B., Czeisler, C.A., Dijk, D.J., Duffy, J.F., Folkard, S., Minors, D., Totterdell, P., Waterhouse, J.: Complex interaction of the sleep-wake cycle and circadian phase modulates mood in healthy subjects. Arch. Gen. Psychiatr. **54**(2), 145–152 (1997)

Bonnet, M.H.: Sleep deprivation. In: Kryger, M.H., Roth, T., Dement, W.C. (eds.) Principles and Practice of Sleep Medicine, 3rd edn, pp. 53–71. WB Saunders, Philadelphia, PA (2000)

Bootzin, R.R.: Stimulus control treatment. Paper presented at the american psychological association annual meeting (1972)

Bootzin, R.R., Perlis, M.L.: Nonpharmacologic treatments of insomnia. J. Clin. Psychiatr. **53**(Suppl), 37–41 (1992)

Broman, J.E., Lundh, L.G., Hetta, J.: Insufficient sleep in the general population. Neurophysiol. Clin. **26**, 30–39 (1996)

Bureau of Labor Statistics: Sports and exercise. http://www.bls.gov/spotlight/2008/sports/ (2008). Accessed 25 Aug 2009

Buysse, D.J., Reynolds, C.F., III,, Monk, T.H., Berman, S.R., Kupfer, D.J.: The Pittsburgh sleep quality index: a new instrument for psychiatric practice and research. Psychiatr. Res. **28**(2), 193–213 (1989)

Cartwright, R.: Dreaming as a mood regulation system. In: Kryger, M.H., Roth, T., Dement, W.C. (eds.) Principles and Practice of Sleep Medicine, 4th edn, pp. 565–572. W.B. Saunders, Philadelphia, PA (2005)
Cohen, D.B.: Toward a theory of dream recall. Psychol. Bull. **81**, 138–154 (1974)
Cole, R.J.: Non-pharmacologic techniques for promoting sleep. Clin. Sports Med. **24**, 343–353 (2005)
Cuartero, P., Estivill, E.: Sleeping, Dreaming and Resting: The Keys to Revitalising Sleep. RBA, Hospes, AIPS, Barcelona (2007)
Czeisler, C.A., Buxton, O.M., Khalsa, S.B.S.: The human circadian timing system and sleep-wake regulation. In: Kryger, M.H., Roth, T., Dement, W.C. (eds.) Principles and Practice of Sleep Medicine, 4th edn, pp. 375–394. W.B. Saunders, Philadelphia, PA (2005)
De Niet, G., Tiemens, B., Lendemeijer, B., Hutschemaekers, G.: Music-assisted relaxation to improve sleep quality: Meta-analysis. J. Adv. Nurs. **65**(7), 1356–1364(9) (2009)
Dinges, D.F., Kribbs, N.B.: Performing while sleepy: effects of experimentally induced sleepiness. In: Monk, T.H. (ed.) Sleep, Sleepiness and Performance, pp. 97–128. Wiley, New York, NY (1991)
Dinges, D.F., Pack, F., Williams, K., Gillen, K.A. et al.: Cumulative sleepiness, mood disturbance, and psychomotor vigilance performance decrements during a week of sleep restricted to 4±5 hours per night. Sleep **20**, 267–277 (1997)
Du, J., Postma, C.E., Van Vugt, H.C., Zwartkruis, E.: Understanding the user needs for sleep enhancement. Philips Internal Report (2008)
Elliott, H.: The use of diaries in sociological research on health experience. Sociol. Res. Online **2**(2), 1–13 (1997)
Epstein, L.J., Mardon, S.: The Harvard Medical School Guide to a Good Night's Sleep. McGraw Hill, New York, NY (2006)
Erickson, T.D., Simon, H.A.: Protocol Analysis: Verbal Reports as Data. The MIT Press, Cambridge (1985)
Evans, D.: The effectiveness of music as an intervention for hospital patients: a systematic review. J. Adv. Nurs. **37**(1), 8–18 (2002)
Fahlström, M., Yeap, J.S., Alfredson, H., Söderman, K.: Shoulder pain – A common problem in world-class badminton players. Scand. J. Med. Sci. Sports **16**(3), 168–173 (2006)
Fitch, T., Armitage, R.: Variations in cognitive style among high and low frequency dream recallers. Personal. Individ. Differ. **10**, 869–875 (1989)
Fontana, A., Frey, J.H.: Interviewing: the art of science. In: Denzin, N., Lincoln, Y. (eds.) Handbook of Qualitative Research, pp. 361–376. Sage, London (1994)
Franzen, P.L., Siegle, G.J., Buysse, D.J.: Relationships between affect, vigilance, and sleepiness following sleep deprivation. J. Sleep Res. **17**, 34–41 (2008)
Garfin, S.R., Pye, S.A.: Bed design and its effect on chronic low back pain – a limited controlled trial. Pain **10**, 87–91 (1981)
George, C.F.: Driving simulators in clinical practice. Sleep Med. Rev. **7**, 311–320 (2003)
Gilchrist, S., Davidson, J., Shakespeare-Finch, J.: Dream emotions, waking emotions, personality characteristics and well-being – a positive psychology approach. Dreaming **17**(3), 172–185 (2007)
Gosselin, N., Lassonde, M., Petit, D., Leclerc, S., Mongrain, V., Collie, A., Montplaisier, J.: Sleep following sport-related concussions. Sleep Med. **10**(1), 35–46 (2009)
Grossman, E., Grossman, A., Schein, M.H., Zimlichman, R., Gavish, B.: Breathing-control lowers blood pressure. J. Hum. Hypertens. **15**, 263–269 (2001)
Haack, M., Mullington, J.M.: Sustained sleep restriction reduces emotional and physical well-being. Pain **199**, 56–64 (2005)
Harvey, A.G., Tang, N.K.Y., Browning, L.: Cognitive approaches to insomnia. Clin. Psychol. Rev. **25**, 593–611 (2005)
Hauri, P., Percy, L., Hellekson, C., Hartmann, E., Russ, D.: The treatment of psychophysiologic insomnia with biofeedback: a replication study. Biofeedback Self Regul. **7**, 223–235 (1982)

Hislop, J., Arber, S., Meadows, R., Venn, S.: Narratives of the night: the use of audio diaries in researching sleep. Sociol. Res. Online **10**, 4 (2005)

Hoddes, E., Zarcone, V., Smythe, H., et al.: Quantification off sleepiness: a new approach. Psychophysiology **10**, 431–436 (1973)

Horne, J., Anderson, C., Platten, C.: Sleep extension versus nap or coffee, within the context of sleep debt. J. Sleep Res. **17**(4), 432–436 (2008)

Horne, J.A., Ostberg, O.: A self-assessment questionnaire to determine morningness-eveningness in human circadian rhythms. Int. J. Chronobiol. **4**, 97–110 (1976)

Jacobson, B.H., Boolani, A., Smith, D.B.: Changes in back pain, sleep quality, and perceived stress after introduction of new bedding systems. J. Chiropr. Med. **8**(1), 1–8 (2009)

Jennum, P., Jensen, R.: Sleep and headache. Sleep Med. Rev. **6**(6), 471–479 (2002)

Johns, M.W.: A new method for measuring daytime sleepiness: the epworth sleepiness scale. Sleep **14**, 540–545 (1991)

Krauchi, K., Cajochen, C., Werth, E.: Physiology – warm feet promote the rapid onset of sleep. Nature **401**(6748), 36–37 (1999)

Krauchi, K., Cajochen, C., Werth, E., Wirz-Justice, A.: Functional link between distal vasodilation and sleep-onset latency? Am. J. Physiol. **278**, R741–R748 (2000)

Liao, W.-C.: Effects of passive body heating on body temperature and sleep regulation in the elderly: a systematic review. Int. J. Nurs. Stud. **39**(8), 803–810 (2002)

Liao, W.-H., Yang, C.M.: Video-based activity and movement pattern analysis in overnight sleep studies. In: Proceedings of the 19th International Conference on Patterns Recognition 2008, pp. 1–4 (2008)

Lichstein, K.L., Riedel, B.W., Means, M.K.: Psychological treatment of late-life insomnia. In: Schulz, R., Maddox, G., Lawton, M.P. (eds.) Annual Review of Gerontology and Geriatrics: Vol. 18. Focus on Interventions Research with Older Adults, pp. 74–110. Springer, New York, NY (1999)

Lomeli, H.A., Olmos-Perez, I., Talero-Gutierrez, C., Moreno, C.B., Gonzalez-Reyes, R., Palacios, L., De la Pena, F., Munoz-Delgado, J.: Sleep evaluation scales and questionnaires: a review. Actas Esp. Psiquiatr. **36**, 50–59 (2008)

Manber, R., Harvey, A.: Historical perspective and future directions in cognitive behavioral therapy for insomnia and behavioral sleep medicine. Clin. Psychol. Rev. **25**(5), 535–538 (2005)

McArdle, N., Kingshott, R., Engleman, H.M., et al.: Partners of patients with sleep apnoea/hypopnoea syndrome: effect of CPAP treatment on sleep quality and quality of life. Thorax **56**, 513–518 (2001)

Monk, T.H., Flaherty, J.F., Frank, E., Hoskinson, K., Kupfer, D.J.: The social rhythm metric. An instrument to quantify the daily rhythms of life. J. Nerv. Ment. Dis. **178**, 120–126 (1990)

Monk, T.H., Reynolds, C.F., Kupfer, D.J., Buysse, D.J., Coble, P.A., Hayes, A.J., Machen, M.A., Petrie, S.R., Ritenour, A.M.: The Pittsburgh sleep diary. J. Sleep Res. **3**, 111–120 (1994)

Morin, C.M., Bootzin, R.R., Buysse, D.J., et al.: Psychological and behavioral treatment of insomnia: update of the recent evidence (1998–2004). Sleep **29**(11), 1398–1414 (2006a)

Morin, C.M., Hauri, P.J., Espie, C.A., et al.: Non-pharmacologic treatment of chronic insomnia. An american academy of sleep medicine review. Sleep **22**(8), 1134–1156 (1999)

Morin, C.M., LeBlanc, M., Daley, M., Gregoire, J.P., Merette, C.: Epidemiology of insomnia: prevalence, self-help treatments, consultations, and determinants of self-seeking behaviours. Sleep Med. **7**, 123–130 (2006b)

Natale, V., Cicogna, P.: Circadian regulation of subjective alertness in morning and evening types. Personal. Individ. Differ. **20**, 491–497 (1996)

National Sleep Foundation: Sleep in America Poll. URL: http://www.sleepfoundation.org/atf/cf/%7Bf6bf2668-a1b4-4fe8-8d1a-a5d39340d9cb%7D/2008%20POLL%20SOF.PDF (2008)

National Sleep Foundation: Depression and sleep. http://www.sleepfoundation.org/article/sleep-topics/depression-and-sleep (2009)

Opp, M.R.: Sleeping to fuel the immune system: mammalian sleep and resistance to parasites. BMC Evol. Biol. **9**(1), 8 (2009)

Pandi-Perumal, S.R., Smits, M., Spence, W., Srinivasan, V., Cardinali, D.P., Lowe, A.D., et al.: Dim light melatonin onset (DLMO): a tool for the analysis of circadian phase in human sleep and chronobiological disorders. Prog. Neuropsychopharmacol. Biol. Psychiatr. **3**, 1–11 (2007)

Pankhurst, F.P., Horne, J.A.: The influence of bed partners on movement during sleep. Sleep **17**, 308–315 (1994)

Preece, J., Rogers, Y., Sharp, H.: Interaction Design: Beyond Human-Computer Interaction. Wiley, New York, NY (2002)

Raymann, R.J.E.M., Swaab, D.F., Van Someren, E.J.W.: Cutaneous warming promotes sleep onset. Am. J. Physiol. Regul. Integr. Comp. Physiol. **288**, R1589–R1597 (2005)

Raymann, R.J.E.M., Swaab, D.F., Van Someren, E.J.W.: Skin temperature and sleep-onset latency: changes with age and insomnia. Physiol. Behav. **90**(2-3), 257–266 (2007)

Robson, C.: Real World Research: A Resource for Social Scientists and Practitioner-Researchers. Blackwell publishing, Oxford (1993)

Roenneberg, T., Kuehnle, T., Juda, M., Kantermann, T., Allebrandt, K., Gordijn, M., Merrow, M.: Epidemiology of the human circadian clock. Sleep Med. Rev. **11**(6), 429–438 (2007)

Roenneberg, T., Wirz-Justice, A., Merrow, M.: Life between clocks: Daily temporal patterns of human chronotypes. J. Biol. Rhythms **18**, 80–90 (2003)

Schredl, M., Atanasova, D., Hormann, K., Maurer, J.T., Hummel, T., Stuck, B.A.: Information processing during sleep: the effect of olfactory stimuli on dream content and dream emotions. J. Sleep Res. **18**(3), 283–290 (2009)

Sleeswijk-Visser, F., Stappers, P.J., Van der Lugt, R., Sanders, E.B.-N.: Contextmapping: experiences from practice. CoDesign **1**(2), 119–149 (2005)

Smith, M.T., Haythornthwaite, J.A.: How do sleep disturbance and chronic pain inter-relate? Insights from the longitudinal and cognitive-behavioral clinical trials literature. Sleep Med. Rev. **8**(2), 119–132 (2004)

Spiegel, K., Leproult, R., Van Cauter, E.: Impact of sleep debt on metabolic and endocrine function. Lancet **354**, 1435–1439 (1999)

Stickgold, R.: Sleep-dependent memory consolidation. Nature **437**(7063), 1272–1278 (2005)

Stone, K.C., Taylor, D.J., McCrae, C.S., Kalsekar, A., Lichsteiri, K.L.: Nonrestorative sleep. Sleep Med. Rev. **12**, 275–288 (2008)

Strawbridge, W.J., Shema, S.J., Roberts, R.E.: Impact of spouses' sleep problems on partners. Sleep **27**(3), 527–531 (2004)

Sung, E.J., Tochihara, Y.: Effects of bathing and hot footbath onsleep in winter. J. Physiol. Anthropol. Appl. Hum. Sci. **19**(1), 21–27 (2000)

Taylor, A., Wright, H., Lack, L.C.: Sleeping-in on the weekend delays circadian phase and increases sleepiness the following week. Sleep Biol. Rhythms **6**(3), 172–179 (2008)

Tryon, W.W.: Issues of validity in actigraphic sleep assessment. Sleep **27**, 158–165 (2004)

Tucker, P., Dahlgren, A., Åkerstedt, T., Waterhouse, J.: The impact of free-time activities on sleep, recovery and well-being. Appl. Ergon. **39**(5), 653–662 (2008)

Urponen, H., Vuori, I., Hasan, J., Partinen, M.: Self-evaluations of factors promoting and disturbing sleep: an epidemiological survey in Finland. Soc. Sci. Med. **26**, 443–450 (1988)

Van Der Werf, Y.D., Altena, E., Schoonheim, M.M., Sanz-Arigita, E.J., Vis, J.C., De Rijke, W., Van Someren, E.J.W.: Sleep benefits subsequent hippocampal functioning. Nat. Neurosci. **12**, 122–123 (2009)

Van Someren, E.J.W.: More than a marker: Interaction between the circadian regulation of temperature and sleep, age-related changes, and treatment possibilities. Chronobiol. Int. **17**, 313–354 (2000)

Van Someren, E.J., Drosopoulos, R.R., Collins, S., et al.: Effect of body temperature manipulation on pulse wave amplitude and sleep onset latency. Sleep **25**(Abstract Suppl), A128 (2002)

Van Vugt, H.C., Du, J., Raymann, R.J.E.M.: Using qualitative methods to study sleep in home contexts. Sleep **32**(Suppl.), 1134 (2009)

Venkatraman, V., Chuah, Y.M.L., Huettel, S.A., Chee, M.W.L.: Sleep deprivation elevates expectation of gains and attenuates response to losses following risky decisions. Sleep **30**, 603–609 (2007)

Venn, S.: 'It's okay for man to snore': The influence of gender on sleep disruption in couples. Sociol. Res. **12**(5) (2007)
Vossen, S., De Ruyter, B., Hart-De Ruijter, E., De Kort, Y., Ysselstein, W., Van Vugt, H.C.: The effects of sleep on mood in a healthy adult sample. Sleep **32**(Suppl.) (2009)
Walker, M.P.: Cognitive consequences of sleep and sleep loss. Sleep Med. **9**(1), S29–S34 (2008)
Walsh, J.K., Dement, W.C., Dinges, D.F.: Sleep medicine, public policy, and public health. In: Kryger, M.H., Roth, T., Dement, W.C. (eds.) Principles and Practice of Sleep Medicine, 4th edn, pp. 648–656. W.B. Saunders, Philadelphia, PA (2005)
Wicklow, A., Espie, C.A.: Intrusive thoughts and their relationship to actigraphic measurement of sleep: towards a cognitive model of insomnia. Behav. Res. Ther. **38**, 679–693 (2000)
Youngstedt, S.D.: Effects of exercise on sleep. Clin. Sports Med. **24**, 355–365 (2005)
Yu, C.K.-C.: Facilitating dream recall in Chinese people. Dreaming **16**(3), 186–195 (2006)

Chapter 10
Telling the Story and Re-Living the Past: How Speech Analysis Can Reveal Emotions in Post-traumatic Stress Disorder (PTSD) Patients

Egon L. van den Broek, Frans van der Sluis, and Ton Dijkstra

Abstract A post-traumatic stress disorder (PTSD) is a severe stress disorder and, as such, a severe handicap in daily life. To this date, its treatment is still a big endeavor for therapists. This chapter discusses an exploration towards automatic assistance in treating patients suffering from PTSD. Such assistance should enable objective and unobtrusive stress measurement, provide decision support on whether or not the level of stress is excessive, and, consequently, be able to aid in its treatment. Speech was chosen as an objective, unobtrusive stress indicator, considering that most therapy sessions are already recorded anyway. Two studies were conducted: a (controlled) stress-provoking story telling (SPS) and a(n ecologically valid) re-living (RL) study, each consisting of a "happy" and an "anxiety triggering" session. In both studies the same 25 PTSD patients participated. The Subjective Unit of Distress (SUD) was determined as a subjective measure, which enabled the validation of derived speech features. For both studies, a Linear Regression Model (LRM) was developed, founded on patients' average acoustic profile. It used five speech features: amplitude, zero crossings, power, high-frequency power, and pitch. From each feature, 13 parameters were derived; hence, in total 65 parameters were calculated. Using LRMs, respectively 83 and 69% of the variance was explained for the SPS and RL study. Moreover, a set of generic speech signal parameters was presented. Together, the models created and parameters identified can serve as the foundation for future artificial therapy assistants.

No laga duele bieu: Skavisábo di nobo.
Let not woes of old enslave you anew.
– Nydia Ecury

E.L. van den Broek (✉)
Human Media Interaction (HMI), University of Twente, 7500 AE, Enschede, The Netherlands; Karakter University Center, Radboud University Medical Center Nijmegen, 6500 HB Nijmegen, The Netherlands
e-mail: vandenbroek@acm.org

J. Westerink et al. (eds.), *Sensing Emotions*, Philips Research Book Series 12,
DOI 10.1007/978-90-481-3258-4_10,

10.1 Introduction

In our modern society, many people experience stress, sometimes for just a brief moment, at other times for prolonged periods of time. Stress can be defined as a feeling of pressure or tension, caused by influences of the outside world. It can be accompanied by positive and by negative feelings. It affects our physical state, for instance by increasing our heart rate and blood pressure, and freeing stress hormones like (nor)adrenaline and (nor)epinephrine (Kosten et al., 1987), stimulating autonomic nerve action. Stress may become harmful if it occurs for too long or too frequently, or if it occurs during a traumatic experience. It may result, for instance, in depression, insomnia, or Post-Traumatic Stress Disorders (PTSD) (Ehlers et al., 2010; L. Mevissen, 2010; Ray, 2008; Rubin et al., 2008). To make it even worse, such stress related disorders stigmatize the people suffering from them, which in itself is an additional stressor (Rüscha et al., 2009a, b).

Depression cannot always be related to a specific cause, though several contributing factors have been identified: e.g., genetic vulnerability and unavoidability of stress (Greden, 2001). More specifically, certain stressful life events (e.g., job loss, widowhood) can lead to a state of depression. Furthermore, chronic role-related stress is significantly associated with chronically depressed mood (Kessler, 1997). Note that the experience of stress is associated with the onset of depression, and not with the symptoms of depression.

Insomnia often has a fairly sudden onset caused by psychological, social, or medical stress (Healey et al., 1981). Nevertheless, in some cases, it may develop gradually and without a clear stressor. Insomnia is characterized by sleep deprivation, and associated with increased physiological, cognitive, or emotional arousal in combination with negative conditioning for sleep (American Psychiatric Association, 2000).

Traumas can originate from a range of situations, such as warfare, natural disaster, and interpersonal violence such as sexual, physical, and emotional abuse, intimate partner violence, or collective violence (e.g., experiencing a bank robbery) (Ray, 2008). In such cases, a posttraumatic stress disorder (PTSD) may arise, which can be characterized by a series of symptoms and causes (Ehlers et al., 2010; L. Mevissen, 2010; Ray, 2008; Rubin et al., 2008), summarized in Table 10.1.

10.2 Post-traumatic Stress Disorder (PTSD)

In our study, we studied the emotions in Post-traumatic Stress Disorder (PTSD) patients, who suffered from Panic Attacks, Agoraphobia, and Panic Disorder with Agoraphobia (L. Mevissen, 2010; Sánchez-Meca et al., 2010).

A Panic Attack is a discrete period in which there is a sudden onset of intense apprehension, fearfulness or terror, often associated with feelings of impending doom. During these Panic Attacks, symptoms such as shortness of breath, palpitations, chest pain or discomfort, choking or smothering sensations, and fear of

Table 10.1 Introduction on (the DSM-IV TR (American Psychiatric Association, 2000) criteria for) posttraumatic stress disorder (PTSD)

Trauma can cause long-term physiological and psychological problems. This has been recognized for centuries. Such suffering (e.g., accompanying a posttraumatic stress disorder, PTSD), can be characterized in terms of series of symptoms and causes. Traumas can originate from a range of situations, either short or long lasting; e.g., warfare, natural disasters such as earthquakes, interpersonal violence such as sexual, physical, and emotional abuse, intimate partner violence, and collective violence

Diagnostic criteria as defined by the DSM-IV TR (American Psychiatric Association, 2000) comprise six categories, each denoting their various indicators:

1. Exposure of the person to a traumatic event
2. Persistent reexperience of the traumatic event
3. Persistent avoidance of stimuli, associated with the trauma, and numbing of general responsiveness (not present before the trauma)
4. Persistent symptoms of increased arousal, not present before the trauma
5. Duration of the disturbance (symptoms in criteria 2, 3, and 4) is more than one month
6. The disturbance causes clinically significant distress or impairment in social, occupational, or other important areas of functioning

Many other symptoms have also been mentioned; e.g., weakness, fatigue, loss of will power, and psychophysiological reactions such as gastrointestinal disturbances. However, these are not included in the DSM-IV TR diagnostic criteria

Additional diagnostic categories are also suggested for victims of prolonged interpersonal trauma, particularly early in life. These concern problems are related to: (1) regulation of affect and impulses, (2) memory and attention, (3) self-perception, (4) interpersonal relations, (5) somatization, and (6) systems of meaning. Taken together, PTSD includes a broad variety of symptoms and diagnostic criteria. Consequently, the diagnosis is hard to make, as is also the case for various other mental disorders

"going crazy" or losing control are present. The Panic Attack has a sudden onset and builds rapidly to a peak (usually in 10 min or less). Panic Attacks can be unexpected (uncued), situationally bound (cued), or situationally predisposed (Sánchez-Meca et al., 2010).

Agoraphobia is anxiety about, or avoidance of, places or situations from which escape might be difficult (or embarrassing), or in which help may not be available in the event of having a panic attack or panic-like symptoms (Sánchez-Meca et al., 2010).

Panic Disorder with Agoraphobia is characterized by both recurrent and unexpected Panic Attacks, followed by at least one month of persistent concern about having another Panic Attack, worries about the possible implications or consequences of such attacks, or a significant behavioral change related to these attacks. The frequency and severity of Panic Attacks vary widely, but Panic Disorder as described here has been found in epidemiological studies throughout the world. Panic Disorders Without and With Agoraphobia are diagnosed two to three times as often in women than in men. The age of onset of Panic Disorders varies considerably, but most typically lies between late adolescence and the mid-thirties. Some individuals may have episodic outbreaks with years of remission in between, and others may have continuous severe symptomatology (Sánchez-Meca et al., 2010).

Due to its large inter-individual variability and its broad variety of symptoms, the diagnosis of PTSD is hard to make (Ehlers et al., 2010; L. Mevissen, 2010; Ray, 2008; Rubin et al., 2008). At the same time, it is clear that an efficient treatment of PTSD requires an objective and early diagnosis of the patients' problems and their therapeutic progress. Assessing the emotional distress of a patient is therefore of the utmost importance. Therapists have developed a range of questionnaires and diagnostic measurement tools for this purpose, e.g., (Knapp and VandeCreek, 1994; Sánchez-Meca et al., 2010). Regrettably, these may be experienced as a burden by clients, because it takes the time and willingness of the clients to complete them.

In addition, several other problems arise when a clinician tries to assess the degree of stress in the patient. First, during the appraisal of a stress response, a stressor may not always be seen as stressful enough to be a cause for the mental illness. In other words, although the client may experience it as hugely stressful, the clinician might not always acknowledge it as such. Second, when measuring the response to a stressor, the clinician may rely on introspection and expertise, but these are always to some extent subjective and they also rely on the communicative abilities, truthfulness, and compliance of the client in question. Third, at times it may not be completely clear which (aspect of) the experienced stressor led to the excessive stress response. Finally, the evaluation of the progress in treatment is complicated by its gradualness and relativity.

10.3 Developing an Artificial Therapy Assistant

Given these considerations, it is abundantly clear why researchers have searched for more objective, unobtrusive ways to measure emotions in patient populations. In other words, in addition to standardizing their professional approaches, therapists have sought for new sorts of therapy evaluation methods that are applicable to real-life situations and measure real emotions.

In our own study, we have made an attempt to develop an artificial therapy assistant for patients with a PTSD in terms of an analysis of characteristics of the speech of such patients during two tasks: their telling of a stress-provoking story, or their verbally reliving of the traumatic event. In addition, we linked the emotions that we measured in these two speech circumstances to those that were reported by the PTSD patients using the more standard measurement of the Subjective Unit of Distress, based on Likert scale questionnaire data.

In the following sections, we will first describe both the story telling and trauma reliving techniques themselves. They provided us with stretches of speech, which we analyzed with respect to a series of signal characteristics to detect emotions. After discussing our speech analysis technique, we will explain how the Subjective Unit of Distress is standardly measured. This will then be followed by a more detailed report of our experimental study. We will end the chapter with an evaluation of our novel approach to stress and emotion measurement.

10.4 Story Telling and Reliving the Past

As described above, the PTSD patients in our study suffered from Panic Attacks. During and directly after a Panic Attack, there usually is a continuous worrying by the client about a new attack, which induces an acute and almost continuous form of stress. In our main studies, we attempted to mimic such stress in two ways; see also Fig. 10.1.

First, in the Stress-Provoking Story (SPS) study, the participants read a stress-provoking or a positive story aloud. Here, story telling was used as the preferred method to elicit true emotions in the patient. This method allows great methodological control over the invoked emotions, in the sense that every patient reads exactly the same story. The fictive stories were constructed in such a way that they would induce certain relevant emotional associations. Thus, by reading the words and understanding the story line, negative or positive associations could be triggered. The complexity and syntactic structure of the different stories were controlled for to exclude the effects of confounding factors. The negative stories were constructed to invoke anxiety, as it is experienced by patients suffering from PTSD. Anxiety is, of course, one of the primary stressful emotions. The positive stories were constructed to invoke a positive feeling of happiness.

Second, in the Re-Living (RL) study, the participants told freely about either their last panic attack or their last joyful occasion. In this study, the participants were asked to tell about the last happy event they could recall, or to re-experience their last panic attack. The therapists assured us that real emotions would be triggered in the reliving sessions with PTSD patients, in particular in reliving the last panic attack.

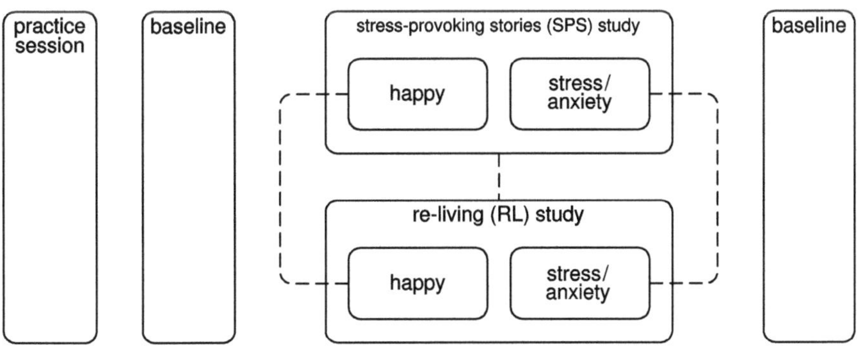

Fig. 10.1 Overview of both the design of the research and the relations (*dotted lines*) investigated. The two studies, SPS and RL, are indicated, each consisting of a happy and a stress/anxiety-inducting session. In addition, baseline measurements were done, before and after the two studies

Because the reliving blocks were expected to have a high impact on the patient's emotional state, a therapist was present for each patient and during all sessions. The two RL sessions were chosen to resemble two phases in therapy: the start and the end of it. Reliving a panic attack resembles the trauma in its full strength, as at the moment of intake of the patient. Telling about the last happy event a patient experienced, resembles a patient who is relaxed or (at least) in a "normal" emotional condition. This should resemble the end of the therapy sessions, when the PTSD has disappeared or is diminished.

10.5 Emotion Detection by Means of Speech Signal Analysis

The emotional state that people are in (during telling a story or reliving the past) can be detected by measuring various signals, such as physiological states, movements, pupil dilation, computer vision techniques, and speech signals. Due to technological developments, some biosignals can now be monitored through ring-like and earring-like devices. However, these and other devices to record biosignals must be attached to a patient's body (van den Broek et al., 2009a, 2010a, b, d).

In our research, we focussed on speech analysis, because this type of analysis is completely unobtrusive (van den Broek et al., 2009a, 2010a, b, d; Zeng et al., 2009). In addition, the communication in therapy sessions is often recorded anyway. Hence, no additional technical effort has to be made on the part of the therapists. Furthermore, because therapy sessions are generally held under controlled conditions in a room shielded from noise, the degree of speech signal distortion is limited.

There is a vast literature on the relationship between speech and emotion. Various speech features have been shown to be sensitive to experienced emotions; see, e.g., (Cowie et al., 2001; Murray and Arnott, 1993; Scherer, 2003; Ververidis and Kotropoulos, 2006; Zeng et al., 2009). In this research, we measured five characteristics of speech:

1. the power (or intensity or energy) of the speech signal; e.g., see Table 10.2 and (Cowie et al., 2001; Murray and Arnott, 1993);
2. its fundamental frequency (F0) or pitch, see also Table 10.2 and (Cowie et al., 2001; Ladd et al., 1985; Murray and Arnott, 1993; Scherer, 2003; Ververidis and Kotropoulos, 2006);
3. the zero-crossings rate (Kedem, 1986; Rothkrantz et al., 2004);
4. its raw amplitude (Murray and Arnott, 1993; Scherer, 2003); and
5. the high-frequency power (Banse and Scherer, 1996; Cowie et al., 2001; Murray and Arnott, 1993; Rothkrantz et al., 2004).

All of these have been considered as useful, for the measurement of experience emotions. Moreover, we expect them to be complementary to a high extent.

Table 10.2 Speech signal analysis: A sample from history

Throughout the previous century, extensive investigations have been conducted on the functional anatomy of the muscles of the larynx; e.g., (Lenneberg, 1967; Hirano et al., 1969). It was shown that when phonation starts, an increase in electrical activity emerges in the laryngeal muscles. Also with respiration, slight electrical activity was found in the laryngeal muscles. These processes are highly complex as speech is an act of large motor complexity, requiring the activity of over 100 muscles (Lenneberg, 1967). These studies helped to understand the mechanisms of the larynx during phonation; cf. (Titze and Hunter, 2007). Moreover, algorithms were developed to extract features (and their parameters) from the human voice. This aided further research towards the mapping of physical features, such as frequency, power, and time, on their psychological counterparts, pitch, loudness, and duration (Cohen and 't Hart, 1967).

In the current research, the physical features are assessed for one specific cause: stress detection. One of the promising features for voice-induced stress detection is the fundamental frequency (F0), which is a core feature in this study. The F0 of speech is defined as the number of openings and closings of the vocal folds per minute, which occurs in a cyclic manner. These cycles are systematically reflected in the electrical impedance of the muscles of the larynx. In particular, the cricothyroid muscle has shown to have a direct relation with all major F0 features (Collier, 1975). In addition, it should be noted that F0 has a relation with another, very important, muscle: the heart. It was shown that the F0 of a sustained vowel is modulated over a time period equal to that of the speaker's heart cycle, illustrating its ability to express one's emotional state (Orlikoff and Baken, 1989).

Through recording of speech signals, its features (e.g., amplitude and F0) can be conveniently determined. This has the advantage that no obtrusive measurement is necessary. Only a microphone, an amplifier, and a recording device are needed. Subsequently, for the determination of F0, appropriate filters (either hardware or software) can increase the relative amplitude of the lowest frequencies and reduce the high- and mid-frequency energy in the signal. The resulting signal contains little energy above the first harmonic. In practice, the energy above the first harmonic is filtered, in a last phase of processing.

Harris and Weiss (Harris and Weiss, 1963) were the first to apply Fourier analysis to compute the F0 from the speech signal. Some alternatives for this approach have been presented in literature; e.g., wavelets (Wendt and Petropulu, 1996). However, the use of Fourier analysis has become the dominant approach. Consequently, various modifications on the original work of Harris and Weiss (Harris and Weiss, 1963) have been applied and various software and hardware pitch extractors were introduced throughout the second half of the twentieth century; e.g., cf. (Dubnowski et al., 1976) and (Rabiner et al., 1976). For the current study, we adopted the approach of Boersma (Boersma, 1993) to determine the F0 of speech.

10.6 The Subjective Unit of Distress (SUD)

To evaluate the quality of our speech analysis, we must compare it to an independent measure of distress. We compared the results of our speech features to those obtained from a standard questionnaire, which measured the Subjective Unit of Distress (SUD). The SUD was introduced by Wolpe in 1958 and has ever since proven itself as a reliable measure of a person's emotional state. The SUD is measured by means of a Likert scale that registers the degree of distress a person experiences at a particular moment in time. In our case, we used a linear scale with a range between 0 and 10 on which the experienced degree of distress could be indicated by a dot or cross. The participants in our study were asked to fill in the SUD test once every minute; therefore, it became routine during the experimental session.

10.7 Design and Procedure

In our study, 25 female Dutch PTSD patients (mean age: 36) participated of their free will. All patients signed an informed consent and all were aware of the tasks included. The experiment began with a practice session, during which the participants learned to speak continuously for longer stretches of time, because during piloting it was noticed that participants had difficulty in doing this. In addition, the practice session offered them the opportunity to become more comfortable with the experimental setting. Next, the main research started, which consisted of two studies and two baseline sessions. The experiment began and ended with the establishment of the baselines, in which speech and SUD were recorded. Between the two baseline blocks, the two studies, the Stress-Provoking Stories (SPS) study and the Re-Living (RL) study, were presented. The two studies were counterbalanced across participants.

The SPS study aimed at triggering two different affective states in the patient. It involved the telling of two stories, which were meant to induce either fear or a neutral feeling. From each of the sessions, three minutes in the middle of the session were used for analysis. The order of the two story sessions was counterbalanced over participants. Both speech and SUD scores (once per minute) were collected.

The RL study also involved two sessions of three minutes. In one of these, the patients were asked to re-experience their last panic attack. In the other, the patients were asked to tell about the last happy event they could recall. Again, the order of sessions was counterbalanced over participants.

With both studies, problems occurred with one patient. In both cases, the data of this patient were omitted from further analysis. Hence, in both conditions, the data of 24 patients were used for further analysis.

10.8 Features Extracted from the Speech Signal

Recording speech was done using a personal computer, a microphone preamplifier, and a microphone. The sample rate of the recordings was 44.1 kHz, mono channel, with a resolution of 16 bits. All recordings were divided in samples of approximately one minute of speech.

Five features were derived from the samples of recorded speech: raw amplitude, power, zero-crossings, high-frequency power, and fundamental frequency; see also Fig. 10.2. Here, we will give a definition of these five features.

The term power is often used interchangeably with energy and intensity. In this chapter, we will follow (Lyons, 2004) in using the term power. For a domain $[0, T]$, the power of the speech signal is defined:

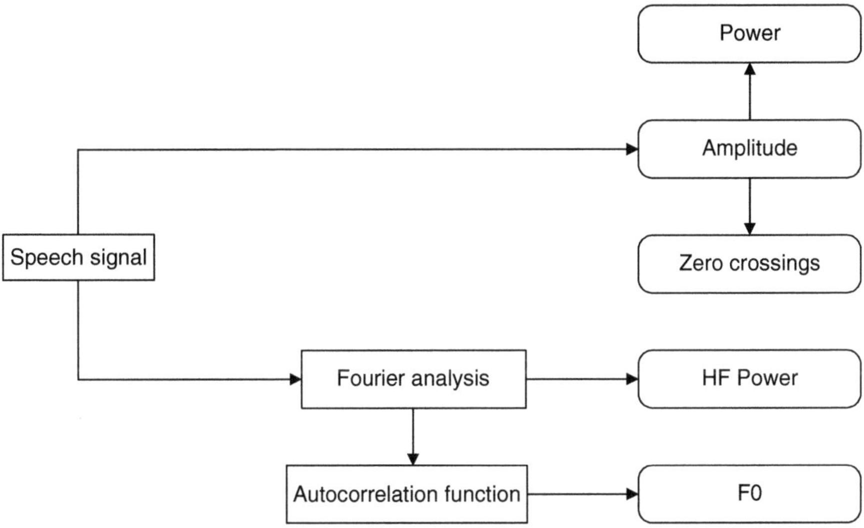

Fig. 10.2 Speech signal processing scheme, as applied in this research. *Abbreviations:* F0: fundamental frequency, HF: high frequency

$$20 \log_{10} \frac{1}{P_0} \sqrt{\frac{1}{T} \int_0^T x^2(t)\, dt}, \tag{1}$$

where the amplitude or sound pressure of the signal is denoted in Pa (Pascal) as $x(t)$ (see also Fig. 10.3a) and the auditory threshold P_0 is $2 \cdot 10^{-5}$ Pa (Boersma and Weenink, 2006).

The power of the speech signal is also described as the Sound Pressure Level (SPL), calculated by the root mean square of the sound pressure, relative to the auditory threshold P_0; i.e., in decibel (dB) (SPL). Its discrete equivalent is defined as (Rienstra and Hirschberg, 2009):

$$20 \log_{10} \frac{1}{P_0} \sqrt{\frac{1}{N} \sum_{n=0}^{N-1} x^2(n)}, \tag{2}$$

where the (sampled) amplitude of the signal is denoted as $x(n)$ in Pa (Pascal) (Boersma and Weenink, 2006). See Fig. 10.3b for an example of the signal power.

The third feature that was computed is the zero-crossings rate of the speech signal. We refrain from defining the continuous model of the zero-crossings rate, since it would require a lengthy introduction and definition; cf. (Rice, 1952). This falls outside the scope of this chapter and does not contribute to its intuitive understanding.

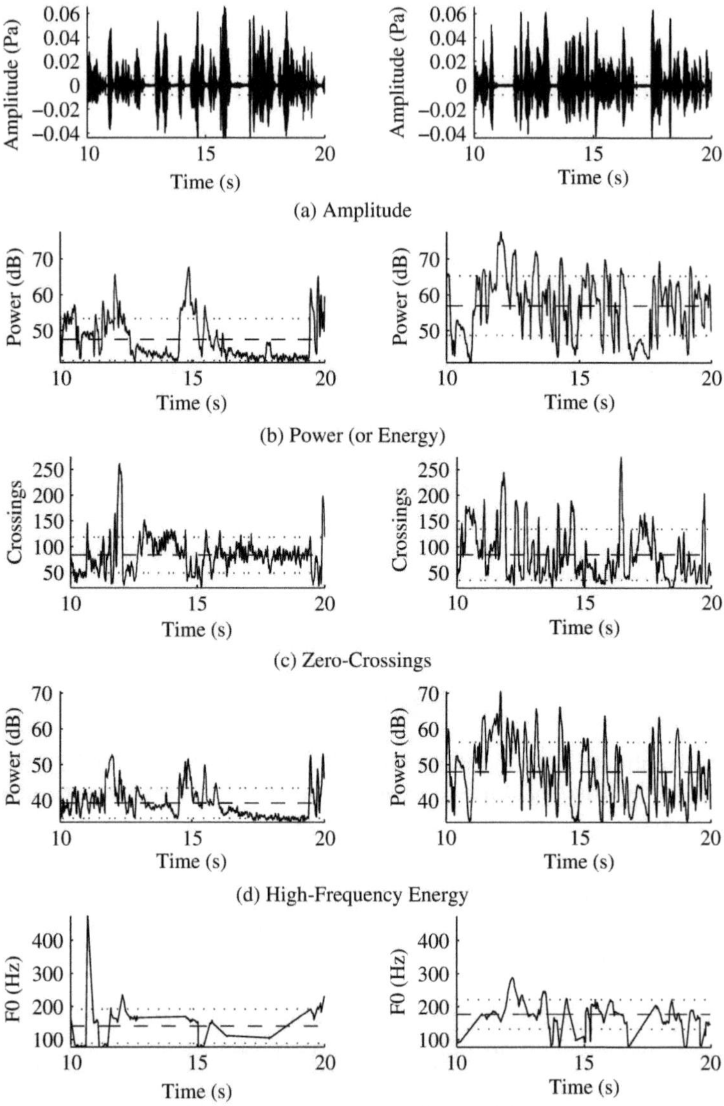

Fig. 10.3 A sample of the speech signal features used of a PTSD patient, who conducted the re-living (RL) study. In each figure, the *middle dotted line* denotes the mean value of the feature. The *upper* and *lower dotted lines* represent one standard deviation from the mean. The SUD scores provided by the patient at the time window of this speech sample were 9 (*left column*) and 5 (*right column*)

Zero crossings can be conveniently defined in a discrete manner, through:

$$\frac{1}{N}\sum_{n=1}^{N-1} \mathbb{I}\{x(n)x(n-1) < 0\}, \tag{3}$$

where N is the number of samples of the signal amplitude x. The $\mathbb{I}\{\alpha\}$ serves as a logical function (Kedem, 1986). An example of this feature is shown in Fig. 10.3c. Note that both power and zero-crossings are defined through the signal's amplitude x; see also Fig. 10.2, which depicts this relation.

The fourth feature that was extracted is the high-frequency power (Banse and Scherer, 1996): the power for the domain $[1000, \infty]$, denoted in Hz. To enable this, the signal was first transformed to the frequency domain; see also Fig. 10.3d. This is done through a Fourier transform $X(f)$ (see also Fig. 10.2), defined as (Lyons, 2004):

$$X(f) = \int_{-\infty}^{\infty} x(t)e^{-j2\pi ft}\, \mathrm{d}t, \tag{4}$$

with j representing the $\sqrt{-1}$ operator. Subsequently, the power for the domain $[F_1, F_2]$ is defined as:

$$20\ \log_{10}\sqrt{\frac{1}{F_2 - F_1}\int_{F_1}^{F_2} |X(f)|^2\, \mathrm{d}t}. \tag{5}$$

For the implementation of the high-frequency power extraction, the discrete Fourier transform (Lyons, 2004) was used:

$$X(m) = \frac{1}{N}\sum_{n=0}^{N-1} x(n)e^{-j2\pi nm/N}, \tag{6}$$

with j representing the $\sqrt{-1}$ operator and where m relates to frequency by $f(m) = mf_s/N$. Here, f_s is the sample frequency and N is the number of bins. The number of bins typically amounts to the next power of 2 for the number of samples being analyzed; e.g., 2,048 for a window of 40 ms. sampled at 44.1 kHz. The power for the domain $[M_1, M_2]$, where $f(M_1) = 1{,}000$ Hz and $f(M_2) = f_s/2$ (i.e., the Nyquist frequency), is defined by:

$$20\ \log_{10}\frac{1}{\mathrm{P}_0}\sqrt{\frac{1}{M_2 - M_1}\sum_{m=M_1}^{M_2} |X(m)|^2}. \tag{7}$$

The fundamental frequency (F0) (or perceived pitch, see Fig. 10.2) was extracted using an autocorrelation function. The autocorrelation of a signal is the cross-correlation of the signal with itself. The cross-correlation denotes the similarity between two signals, as a function of a time-lag between them. In its continuous form, the autocorrelation r of signal x at time lag τ can be defined as (Boersma, 1993):

$$r_x(\tau) = \int_{-\infty}^{\infty} x(t)x(t+\tau)\,\mathrm{d}t \tag{8}$$

In the discrete representation of Eq. (8), the autocorrelation R of signal x at time lag m is defined as (Shimamura and Kobayashi, 2001):

$$R_x(m) = \sum_{n=0}^{N-1} x(n)x(n+m) \tag{9}$$

where N is the length of the signal. The autocorrelation is then computed for each time lag m over the domain $M_1 = 0$ and $M_2 = N - 1$. The global maximum of this method is at lag 0. The local maximum beyond 0, lag m_{max}, represents the F0, if its normalized local maximum $R_x(m_{max})/R_x(0)$ (its harmonic strength) is large enough (e.g., > 0.45). The F0 is derived by $1/m_{max}$. See Fig. 10.3e for an illustrative output of this method.

Throughout the years, various implementations have been proposed for F0 extraction; e.g., (Boersma, 1993; Shimamura and Kobayashi, 2001). See Table 10.2 for a discussion on speech signal processing and on F0 extraction in particular. In this research, we have adopted the implementation as described in (Boersma, 1993). This implementation applies a fast Fourier transform (see also Eqs. (4) and (6)) to calculate the autocorrelation, as is often done; see (Boersma, 1993; Shimamura and Kobayashi, 2001) and Table 10.2. For a more detailed description of this implementation, we refer to (Boersma, 1993).

Of all five speech signal features, 13 statistical parameters were derived: *mean*, *median*, standard deviation (*std*), variance (*var*), minimum value (*min*), maximum value (*max*), range (*max-min*), the quantiles at 10%($q10$), 90%($q90$), 25%($q25$), and 75%($q75$), the inter-quantile-range 10–90% ($iqr10$, $q90$–$q10$), and the inter-quantile-range 25–75% ($iqr25$, $q75$–$q25$). Except for the feature amplitude, the features and statistical parameters were computed over a time window of 40 ms, using a step length of 10 ms; i.e., computing each feature every 10 ms over the next 40 ms of the signal. Hence, in total 65 (i.e., 5×13) parameters were determined from the five speech signal features.

10.9 Results

We separately analyzed the Stress-Provoking Story study and the Re-Living study. The analyses were the same for both studies; with both studies, the SUD scores were reviewed and an acoustic profile was generated.

The acoustic profiles were created with an LRM (Harrell, Jr., 2001). For more information on LRM, we refer to Appendix 1. It was expected that the acoustic profiles would benefit from a range of parameters derived from the five features, as it is known that various features and their parameters have independent contributions to the speech signal (Ladd et al., 1985). In order to create a powerful LRM, backward elimination/selection was applied to reduce the number of predictors. With backward elimination/selection, first all relevant features/parameters are added as predictors to the model (the so-called enter method), followed by multiple iterations removing each predictor for which $p < \alpha$ does not hold (Derksen and Keselman, 1992; Harrell, Jr., 2001). In this research, we chose $\alpha = 0.1$, as the threshold for determining whether or not a variable had a significant contribution to predicting subjective stress.

The backward elimination/selection stops when for all remaining predictors in the model, $p < \alpha$ is true. As the backward method uses the relative contribution to the *model* as selection criteria, the interdependency of the features is taking into account as well. This makes it a robust method for selecting the most relevant features and their parameters. This is crucial for creating a strong model, because it has been shown that inclusion of too many features can reduce the power of a model (Dash and Liu, 1997). Because the general practice of reporting the explained variance of a regression model, R^2, does not take this into account, the adjusted R^2, $\overline{R}^2$ was computed as well. The $\overline{R}^2$ penalizes the addition of extra predictors to the model, and, therefore, is always equal to or lower than R^2.

10.9.1 Results of the Stress-Provoking Story (SPS) Sessions

First, changes with respect to the SUD in the course of the sessions of the SPS study were analyzed. In an Analysis of Variance (ANOVA), no main effects of the SPS session (happy or anxious) or measurement moment (first, second, or third minute of story telling) on the SUD scores were found, nor did any significant interaction effect between these factors appear. A closer look at the SUD scores in the fear session showed that the experienced fear reported by the patients increased in the course of story telling, as indicated by a trend in the ANOVA for the factor measurement moment, $F(2, 67) = 2.59, p < 0.010$. Figure 10.4 illustrates this trend. In addition, Fig. 10.4 shows the confidence intervals, only without variability associated with between-subjects variance; cf. (Cousineau, 2005).

Next, a robust acoustic profile was created of the speech characteristics sensitive to stress. Table 10.4 in Appendix 2 provides the acoustic profile with all significant acoustic features. This profile was generated after 20 iterations of the backward method, leaving 30 significant predictors explaining 81.00% of variance: $R^2 = 0.810, \overline{R}^2 = 0.757, F(30, 109) = 15.447, p < 0.001$. Before applying the backward method (i.e., before any predictors were removed), 50 predictors explained 82.60% of variance: $R^2 = 0.826, \overline{R}^2 = 0.728, F(50, 89) = 8.445$, $p < 0.001$. These results indicate that the amount of variance explained through the acoustic profile is high, as was expected based on literature (Ladd et al., 1985).

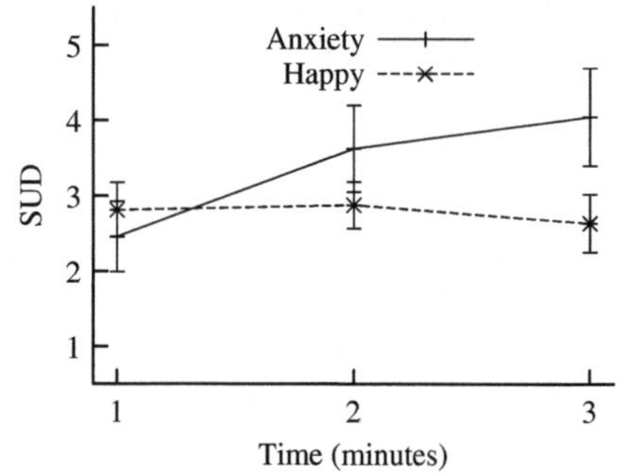

Fig. 10.4 Reported stress over time per session (i.e., anxiety triggering and happy) for the Stress-Provoking Stories (SPS) study

10.9.2 Results of the Re-Living (RL) Sessions

Similar to the analyses performed for the SPS sessions, the analyses for the RL sessions start with an ANOVA of the changes in SUD during the course of the sessions. The results were similar to the SPS analyses: no main effects of the RL session (happy or anxious) or time (first, second, or third minute of story telling) on the SUD scores were found, nor did a significant interaction effect appear. Again, there was a trend in the anxiety triggering condition for patients to report more experience stress later-on in the course of re-living, as indicated by a trend in the ANOVA for the factor time, $F(2,69) = 2.69, p < 0.010$. This trend is also evident in Fig. 10.5. Note that Fig. 10.5 shows the confidence intervals without between-subjects variance; cf. (Cousineau, 2005).

A strong acoustic profile for the RL session was created by means of the speech characteristics that are sensitive to stress. An LRM based upon all relevant features and their parameters (49 predictors) explained 69.10% of variance: $R^2 = 0.691, \overline{R}^2 = 0.530, F(49,94) = 4.29, p < 0.001$. A smaller LRM, based only on the significant features, used 23 predictors explaining 64.80% of variance: $R^2 = 0.648, \overline{R}^2 = 0.584, F(22,121) = 10.12, p < 0.001$. Table 10.5 in Appendix 2

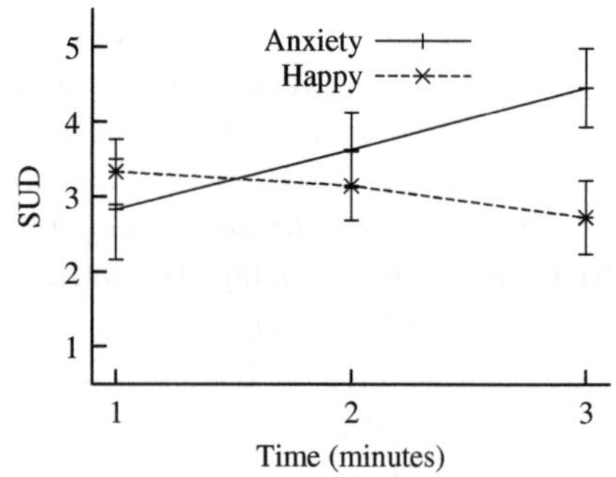

Fig. 10.5 Reported stress over time per session (i.e., anxiety triggering and happy) for the Re-Living (RL) study

shows this acoustic profile with all its significant acoustic features. These results indicate that, for the RL sessions, the subjectively reported stress could be explained very well, as was expected based on literature (Ladd et al., 1985). However, the explained variance was lower than for the SPS sessions.

10.9.2.1 Overview of the Features

A comparison of the LRM of the RL sessions and the SPS sessions shows there are 13 shared predictors: pitch iqr25 and var; amplitude q75, var, and std; power iqr25, q25, and std; zero-crossings q25 and q10; high-frequency power var, std, and mean. However, this comparison is misleading due to the role of the interdependency of the predictors in specifying whether or not it has a significant contribution to the estimate. Hence, for a more appropriate comparison, we used a simpler approach; namely, by computing the linear correlation of each feature and its parameters independent of each other for both data sets (i.e., the RL and SPS data). See Table 10.3 for the results.

Table 10.3 shows which predictors are robust for both data sets and which are not; i.e., which features show a significant linear correlation for the RL as well as the SPS sessions. The F0 is uniformly robust, namely on its mean and cumulative distribution (q10, q25, median, q75, q90). Power and high-frequency power show similar patterns, though more towards parameters describing the lower part of the cumulative distribution (q10, iqr10, iqr25) and more general statistical parameters used to describe the distribution (std, var, range), only without the mean. There is a perfect similarity between power and high-frequency power in which parameters are relevant for both data sets. The features amplitude and zero-crossings have no parameters relevant for both data sets. Concluding, it seems that especially F0, power, and high-frequency power, are robust features for both data sets.

10.10 Discussion

In this section, we will first briefly discuss the results of both the SPS and RL studies. Next, the results on both studies will be compared to each other. Moreover, the results of both studies will be related to relevant theory.

10.10.1 Stress-Provoking Stories (SPS) Study

Using the telling of a carefully created story to induce an affective state, stress was successfully induced in and reported by our PTSD patients. By comparing speech characteristics to a subjective report of stress, we were able to define and evaluate an acoustic profile of stress features in speech. The acoustic profile was shown to explain at best 82.60% of variance of subjectively reported experienced stress.

Table 10.3 Correlations between Subjective Unit of Distress (SUD) and the parameters of the five features derived from the speech signal, both for the Re-Living (RL) and the Stress-Provoking Stories (SPS) study

	Amplitude		Power		ZC		HFP		F0	
Parameter	RL	SPS	RL	SPS	RL	SPS	RL	SPS	RL	SPS
iqr25			$-0.314^{\ddagger}$	$-0.426^{\ddagger}$	$-0.233^{\dagger}$		$-0.327^{\ddagger}$	$-0.355^{\ddagger}$		
q75				$-0.227^{\dagger}$	$-0.258^{\dagger}$			-0.196^{*}	$-0.298^{\ddagger}$	-0.182^{*}
q25									$-0.234^{\dagger}$	$-0.244^{\dagger}$
iqr10		$-0.218^{\dagger}$	$-0.358^{\ddagger}$	$-0.428^{\ddagger}$	-0.197^{*}		$-0.356^{\ddagger}$	$-0.422^{\ddagger}$		
q90		-0.209^{*}		-0.191^{*}	$-0.228^{\dagger}$			-0.189^{*}	$-0.296^{\ddagger}$	$-0.224^{\dagger}$
q10		$0.225^{\dagger}$	0.200^{*}	0.180^{*}		$-0.229^{\dagger}$	$0.222^{\dagger}$	0.168^{*}	$-0.306^{\ddagger}$	-0.193^{*}
median					-0.180^{*}			-0.180^{*}	$-0.271^{\dagger}$	-0.202^{*}
min			$0.223^{\dagger}$			$-0.329^{\ddagger}$	$0.227^{\dagger}$			
max					-0.192^{*}			-0.168^{*}		
range			$-0.282^{\ddagger}$	$-0.243^{\dagger}$	-0.179^{*}		$-0.304^{\ddagger}$	$-0.312^{\ddagger}$		
var		-0.184^{*}	$-0.327^{\ddagger}$	$-0.411^{\ddagger}$	$-0.249^{\dagger}$		$-0.317^{\ddagger}$	$-0.384^{\ddagger}$		
std		-0.202^{*}	$-0.351^{\ddagger}$	$-0.433^{\ddagger}$	$-0.250^{\dagger}$		$-0.354^{\ddagger}$	$-0.413^{\ddagger}$		
mean					$-0.290^{\ddagger}$				$-0.335^{\ddagger}$	$-0.255^{\dagger}$

Levels of significance. $^{*}p < 0.05$, $^{\dagger}p < 0.01$, $^{\ddagger}p < 0.001$.
Abbreviations. ZC: Zero-Crossings rate, HFP: High-Frequency Power.

In interpreting the results, two factors will be differentiated: the experienced and the expressed emotions. In essence, the experienced emotions were targeted by the SUD. Although there was quite some substantial variability in the reported experience, the SUD seemed to have uncovered some expected effects; e.g., the stress in the fear inducing story appeared to develop through the course of telling the story. The substantial variability can be considered a good thing as well, as it might hint at inter-personal differences which were not evidently expected from the highly standardized stimuli, but which the SUD was able to measure; cf. (Lacey et al., 1953; Lacey, 1967). Furthermore, another issue can be noted in the experience of the stories; namely, stories develop over time, which implies that a build-up is necessary before an affective state is induced.

As indicated by the explained variance of the acoustic profile, the expressed emotions seem to reflect the experienced emotions very well. In other words, using triangulation through various speech characteristics and the SUD scores indicated that true emotions were indeed triggered and expressed. Hence, although story telling is only one of many ways to induce emotions, it was particularly useful in creating an emotion-induced speech signal. Contrary to many other methods, this method is likely to have created true emotions.

10.10.2 Re-Living (RL) Study

Apart from the Stress-Provoking Story (SPS) study, our research included a study in which participants re-lived their traumatic event. As such, this research presents unique data, containing very rare displays of intense, real, emotions; hence, a data set with high ecological validity.

Using the RL data set, a Linear Regression Model (LRM) was generated which explained at best 69.10% of variance in SUD scores. Although lower than in the SPS study, it is still a very high percentage of explained variance. In interpreting these results, again, we differentiate between the experienced and expressed emotion and used the SUD scores to capture the experienced emotions. The same issues can be denoted as for the SPS study: the SUD scores tended to vary quite substantial across patients, and both showed a build-up in affective state throughout the session. Hence, the experienced emotions varied between patients, which can be expected as the sessions were relatively less standardized (Lacey et al., 1953; Lacey, 1967); i.e., the patients were merely guided in experiencing true emotions. Furthermore, the latter issue is in line with what is known on emotions and their accompanying reactions; that emotions can (indeed) accumulate over time (Geenen and van de Vijver, 1993; van den Broek and Westerink, 2009).

The expressed emotions are intense displays of emotions; as such, parts of the speech signal even had to be cleaned from non-speech expressions (e.g., crying). Hence, the speech signal clearly reflected emotions. As such, the presented LRM is a rare and clear acoustic profile of true emotions.

10.10.3 Stress-Provoking Stories (SPS) Versus Re-Living (RL)

In comparing the studies, several differences were found: the SUD scores for the RL sessions were not significantly higher than for the SPS sessions, and the explained variance of the acoustic profiles was 13.50% lower for the RL study than for the SPS study. Moreover, when comparing the features by their simple linear correlation with the SUD data, it showed that some features were clearly robust for both studies (i.e., power, high-frequency power, and F0), whereas some were not (i.e., amplitude and zero-crossings rate). In sum, there were 22 parameters (of which 17 were in the amplitude and zero-crossings rate features) which worked for only one of the data sets and 18 parameters which worked for both data sets. The robust parameters could be grouped into specific meaningful parts of the features: for the F0 its mean and cumulative distribution (q10, q25, median, q75, q90), and for power and high-frequency power their lower part of the cumulative distribution (q10, iqr10, iqr25) and more general statistical parameters used to describe the variation of the distribution (std, var, range). Concluding, there were substantial similarities as well as differences between the studies, which will be discussed next.

Considering the experienced emotions, the results were counter-intuitive: The reported stress was not significantly higher in the RL study than in the SPS study. Hence, either the experience was indeed not different from the SPS studies, or introspection is fallible. There were, of course, differences in the experienced emotions between the studies; i.e, the stimuli were different. Story telling was used as a highly standardized laboratory method, whereas the re-living sessions were indeed closer to the patient's experience. Moreover, this view is also supported by the differences between the acoustic profiles and, by qualitative judgements of the patient's psychiatrists also present during the studies. Hence, this would indicate that the SUD scores were a non-perfect mapping on the truly experienced stress. Even if the actual experienced emotions differed between studies, this should not have caused any differences, as the SUD was designed to query this exact experience. Hence, introspection seems to be fallible. Of course, the problems with introspection are not new; tackling them is one of the core motivations for this study. Moreover, we analyzed the SUD scores as an interval scale, an assumption that might not be correct.

The differences between the SPS and the RL study can also be explained by the notion of emotion specifity or cognitive versus emotional stress (Lacey, 1967; Lively et al., 1993; Rüscha et al., 2009a, b). Cognitive stress is defined as the information processing load placed on the human operator while performing a task. Emotional stress is the psychological and physiological arousal due to emotions triggered before or during a task. Both the research setting and the therapeutic setting could have caused cognitive stress; so, this would not discriminate between both studies. However, possibly the cognitive stress had a higher impact on the speech signal obtained with the SPS study than on that obtained with the RL study, where emotional stress was dominant.

Part of the explanation may also be at the expression of emotions. Already more than a century ago (Marty, 1908), the differentiation between emotional and emotive communication was noted. Emotional communication is a type of spontaneous, unintentional leakage or bursting out of emotion in speech. In contrast, emotive communication has no automatic or necessary relation to "real" inner affective states. Emotive communication is a strategic signaling of affective information in speaking to interaction partners that is widespread in interactions. It uses signal patterns that differ strongly from spontaneous, emotional expressions and can be both intentionally and unintentionally accessed (Banse and Scherer, 1996). It is plausible that in the RL study relatively more emotional communication took place, while emotional expressions in the SPS study were based more on features of emotive communication.

When the differences in results between the SPS and the RL study are explained in terms of the distinction between emotional and emotive communication (Banse and Scherer, 1996; Khalil, 2006; Marty, 1908), interesting conclusions can be drawn. The intersection of the parameter sets of both studies should then reflect the aspects of the speech signal that are used in emotional communication. The RL study triggered "real" emotions and in the SPS study probably also "real" emotions were revealed in addition to the emotive communication. Consequently, the parameters unique for the SPS study should reflect characteristics of the speech signal that represent emotive communication. Additionally, the parameters unique for the RL study should reflect characteristics of the speech signal that represent emotional communication. Further research investigating this hypothesis is desirable.

Having discussed hypotheses based on both the distinction between cognitive and emotional stress and the theory on emotive and emotional communication, both notions should also be taken together. Communication as expressed with emotional stress (Lacey, 1967; Lively et al., 1993; Rüscha et al., 2009a, b) and emotional communication (Banse and Scherer, 1996; Marty, 1908) could point to the same underlying construct of emotionally loaded communication. However, this does not hold for cognitive stress (Lacey, 1967; Lively et al., 1993; Rüscha et al., 2009a, b) and emotive communication (Banse and Scherer, 1996; Khalil, 2006; Marty, 1908). It is possible that both cognitive stress and emotive communication have played a significant role in the SPS study. This would then involve a complex, unknown interaction.

10.11 Reflection: Methodological Issues and Suggestions

The design of this research makes it unique in its kind; see also Fig. 10.1. Two studies were conducted, which were both alike and at the same time completely different. The Stress-Provoking Stories (SPS) study comprised a controlled experimental method intended to elicit both stress and more happy feelings. Within the Re-Living (RL) study, true emotions linked to personally experienced situations were facilitated. In both studies the same patients participated. The studies were executed sequentially, in a counterbalanced order.

A question which is often posed is whether "true" emotions can be triggered in controlled research environments. Moreover, if emotions can be triggered in controlled research, how do they relate to emotions experienced in everyday life? Is it only the intensity in which they differ or do different processes underly real-life situations? These questions are hard to answer solely based on a review of literature. Problems arise when one compares empirical studies. Recently, a set of prerequisites for affective signal processing (ASP) have been presented (van den Broek et al., 2009a; van den Broek et al., 2010a, b, d). Although these prerequisites were introduced as guidelines to process biosignals it is posed that they also hold for speech signals, computer vision techniques, and brain-computer interfaces that aim to determine emotions (van den Broek et al., 2010c, 2010).

In total 10 prerequisites for ASP have been proposed: (i) validation (e.g., mapping of constructs on signals), (ii) triangulation, (iii) a physiology-driven approach, (iv) contributions of the signal processing community, (v) identification of users, (vi) temporal construction, (vii) theoretical specification, (viii) integration of biosignals, (ix) physical characteristics, and (x) reflection: a historical perspective (van den Broek et al., 2009a, 2010a, b, d). These will serve as guidelines in our discussion on the pros and cons of this research.

The validity of the current research is high. Content validity is high as (a) the research aimed at a specific group of patients, (b) the SUD as well as the speech signal features and their parameters are chosen with care, all denoted repeatedly in literature; see also Section 10.5, and (c) the SUD in combination with the speech signal features chosen provide a complete image of the patients emotional state, as has been shown. Criteria-related validity is also high as speech was the preferred measurement, being robust and unobtrusive. Moreover, we were able to record emotions real-time. The SUD was provided each minute, which can also be considered as accurate, given the context. The construct validity is limited since for both stress and emotions various definitions exist and no general consensus is present. Moreover, no relations are drawn between emotion, stress, psychological changes, physiological changes, and the speech signal. The ecological validity is high, at least for one of both studies. For the other study the ecological validity is limited, as illustrated by the difference in results between both studies.

The principle of triangulation is applied; that is, multiple operationalizations of constructs were used. The distinct speech signal features could be validated against each other and against the SUD. Extrapolations were made using the data sets of both studies and a set of common discriminating speech features have been identified. Moreover, the SUD was used as ground truth. However, this required introspection of the patients, which is generally not considered as the most reliable measure.

This research did not have the specific aim to employ a physiology-driven approach. However, in practice it could be baptized as such. Solely the speech signal is needed to enable automatic therapy assistance.

Various contributions from the signal processing community have been incorporated in the current research, as is denoted in Section 10.5 and in Table 10.2. However, much more expertise from the signal processing community could be employed with follow-up research.

The users of the envisioned artificial therapy assistant were clearly defined: therapists that treat patients suffering from a PTSD. Possibly, the models developed and features identified in this research can also show their use for other groups of patients.

For speech signal processing the temporal construction is not as crucial as it is for biosignal processing. Due to their nature, various biosignals have different delays (van den Broek et al., 2010a, b, d). In contrast, the speech signal hardly suffers from a delay and all its features have the same latency.

The theoretical specification of the features used is mentioned; e.g., see Sections 10.5 and 10.6, and Table 10.2. However, this chapter did not have the aim to elaborate exhaustively on this issue. Hence, as such, the theoretical specification provided here is limited.

This research has used one signal; hence, no integration of signals have been applied. However, for both studies, the features and their parameters were all integrated in one LRM. Additional other signals were omitted on purpose since they could contaminate the ecological validity of the research, as they would interfere with the actual tasks the patients had to perform.

This chapter did not specify any physical characteristics, as they are of little interest. The artificial therapy assistant should be able to function in a setting as in which this research was conducted; hence, having the same physical characteristics. In general, these are average office settings. Within reason, the speech signal processing scheme 2 should be able to handle changing physical characteristics of an office, which could influence the room's acoustics. However, there are no indications for any problems that could occur as a results of this.

Throughout the chapter, repeatedly, an historical perspective is taken into account. Table 10.2 is even devoted to this. Moreover, for the various speech signals (see Section 10.5) and for the SUD (see Section 10.6), their origins are denoted.

The list of prerequisites is probably not complete. However, it provides an indication for the quality of the methodological foundation of this research and shows where there is room for improvement.

10.12 Conclusions

This chapter has presented two studies in which the same Post-Traumatic Stress Disorder (PTSD) patients participated. This provided us with two unique data sets. Moreover, these data sets could be compared with each other and the influence of SPS and RL could be compared, because except for the task (i.e., resp. story telling and re-living) both studies were the same. This has revealed interesting common denominators as well as differences between both studies, which are of concern for several theoretical frameworks. Moreover, a thorough discussion has been presented, in two phases. First, the results of both studies were discussed and, subsequently, related to each other. Second, a range of aspects concerning the complete research were discussed, using a set of 10 prerequisites. This emphasized the strength of the research presented and also provided interesting pointers for follow-up research.

Derived from the data of each of the studies, a Linear Regression Model (LRM) was developed. These LRMs explained respectively 83% of the variance for the SPS study and 69% of the variance for the RL study, which is both high. Founded on the results of both studies, a set of generic features has been defined; see also Table 10.3. This set could serve as the foundation for the development of models that enable stress identification in a robust and generic manner.

It would be of interest to apply such a model also on patients suffering from other related psychiatric disorders, such as depression (Kessler, 1997; American Psychiatric Association, 2000) and insomnia (Healey et al., 1981; American Psychiatric Association, 2000). Probably, for even less related psychiatric disorders, the current approach would be a good starting point. In such a case, the general framework and speech signal processing scheme (see Fig. 10.2), as presented in this chapter, could be employed. Most likely, only the set of parameters used for the LRM should have to be tailored to the specific disorders.

The speech signal processing approach used in this research could also be linked to approaches that measure physiological responsiveness of PTSD in other ways; e.g., using biosignals (van den Broek et al., 2009a; 2010a, b) or computer vision techniques (Cowie et al., 2001; Zeng et al., 2009). This would facilitate a triangulation of the construct under investigation, providing even more reliable results (van den Broek et al., 2009a). Furthermore, more specific analyses can be conducted, for example, in terms of either the valence and arousal model or discrete emotion categories (van den Broek et al., 2009a). However, it also has its disadvantages, as discussed in the previous section.

The models developed and features and their parameters identified in this research could also be of use for other application areas than psychiatry. It has already been posed that consumer electronics (van den Broek and Westerink, 2009) and artificial intelligence (Picard, 1997) could benefit from (unobtrusive) emotion detection. But also ambient intelligence (van den Broek et al., 2009b), man-machine interaction (van den Broek et al., 2010c), and robotics (Rani et al., 2002; van den Broek, 2010) will certainly benefit from the introduction of such techniques. However, in these cases, the group of people to be analyzed is even more diverse. Hence, obtaining robust results in such settings would be even more challenging than was the case with the current research.

Apart from being unobtrusive, the speech signal processing approach, as applied in the current research, has another major advantage. It enables the remote determination of people's emotional state. This feature enables its use in yet another range of contexts; for example, in telepsychiatry (Hilty et al., 2004) and call-centers (Morrison et al., 2007) that frequently have to cope with highly agitated customers. However, as with the different psychiatric disorders and the other application areas mentioned, also in this case the LRM should be adapted to this task.

Taken together, an important and significant step has been made towards a artificial therapy assistant for treatment of patients suffering from a PTSD in particular and stress-related psychiatric disorders in general. Through the design of the research, it was made sure that "real" emotions were measured. Subsequently, their objective measurement through speech signal processing was shown to be feasible. Models were constructed, founded on a selection from 65 parameters of five speech features. With up to 83% explained variance, the models showed to provide reliable, robust classification of stress. As such, the foundation was developed for an objective, easily usable, unobtrusive, and powerful artificial therapy assistant.

Acknowledgements Most important, we gratefully acknowledge the voluntary participation of the post-traumatic stress disorder (PTSD) patients in this research. Further, we acknowledge Paul Boersma (Institute of Phonetic Sciences, University of Amsterdam, NL) for his work on the program Praat and for his helpful comments in the initial phase of this project. Bert Cranen (Dept. of Language and Speech, Radboud University Nijmegen, NL) is acknowledged for his comments, in particular on Table 10.2.

Appendix 1: Introduction on Linear Regression Models

A linear regression model (LRM) is an optimal linear model of the relationship between one dependent variable (e.g., the SUD) and several independent variables (e.g., the speech features). A linear regression model typically takes the following form:

$$y = \beta_0 + \beta_1 x_1 + \cdots + \beta_p x_p + \varepsilon, \tag{10}$$

where ε represents unobserved random noise, and p represents the number of predictors (i.e., independent variables x and regression coefficients β). The linear regression equation is the result of a linear regression analysis, which aims to solve the following n equations in an optimal fashion:

$$\begin{pmatrix} y_1 \\ y_2 \\ \vdots \\ y_n \end{pmatrix} = \begin{pmatrix} x_{11} & \cdots & x_{1p} \\ x_{21} & \cdots & x_{2p} \\ \vdots & \ddots & \vdots \\ x_{n1} & \cdots & x_{np} \end{pmatrix} \times \begin{pmatrix} \beta_1 \\ \beta_2 \\ \vdots \\ \beta_p \end{pmatrix} + \begin{pmatrix} \varepsilon_1 \\ \varepsilon_2 \\ \vdots \\ \varepsilon_n \end{pmatrix}. \tag{11}$$

Here, there are n equations for n data points of y. As there is normally more than one solution to the problem, a least squares method is used to give the optimal solution. Please consult a handbook (e.g., (Harrell, Jr., 2001)) for more information on the least squares method and its alternatives. A discussion of this topic falls beyond the scope of this chapter.

The following characteristics are used to describe an LRM:

1. Intercept: the value of β_0.
2. Beta (B) and Standard Error (SE): the regression coefficients and standard error of its estimates.
3. Standardized B (β): the standardized Betas, in units of standard deviation of its estimates.
4. T-test (t): a t-test for the impact of the predictor.
5. F-test (F): an ANOVA testing the goodness of fit of the model for predicting the dependent variable.

6. R-square (R^2): the amount of explained variance by the model relative to the total variance in the dependent variable.
7. Adjusted R-square ($\overline{R}^2$): R-square (R^2) penalized for the number of predictors used.

Appendix 2: Specification of the Linear Regression Models Developed

For both the Stress-Provoking Stories (SPS) study and the Re-Living (RL) study, a LRM was developed. To enable, the replication of these LRM, this Appendix provides their specifications. Table 10.4 denotes the LRM of the SPS study. Table 10.5 denotes the LRM of the RL study.

Table 10.4 Linear regression model predicting Subjective Unit of Distress (SUD) for Stress-Provoking Stories (SPS) study

Feature	Parameter	B	SE (B)	β	*t*	*p*
Intercept		0.85	20.50		0.04	= 0.967
Pitch	iqr25	0.12	0.03	0.52	3.77	< 0.001
Pitch	q25	0.08	0.03	0.80	3.09	= 0.003
Pitch	iqr10	−0.11	0.02	−1.07	−4.36	< 0.001
Pitch	q10	−0.11	0.03	−1.17	−4.20	< 0.001
Pitch	min	0.05	0.01	0.21	3.86	< 0.001
Pitch	var	0.00	0.00	1.10	2.44	= 0.016
Pitch	std	−0.24	0.11	−0.98	−2.16	= 0.033
Amplitude	q75	1354.02	238.41	1.82	5.68	< 0.001
Amplitude	q25	1510.39	222.65	2.74	6.78	< 0.001
Amplitude	var	2288.04	552.51	3.47	4.14	< 0.001
Amplitude	std	−329.34	117.01	−3.58	−2.81	= 0.006
Amplitude	mean	18019.42	8023.76	0.17	2.25	= 0.027
Power	iqr25	−1.54	0.42	−2.14	−3.66	< 0.001
Power	q25	−1.92	0.47	−3.63	−4.08	< 0.001
Power	range	0.19	0.08	0.30	2.40	= 0.018
Power	var	−0.45	0.11	−5.58	−4.15	< 0.001
Power	std	11.43	2.16	6.56	5.30	< 0.001
Power	mean	1.93	0.76	3.46	2.53	= 0.013
Zero crossings	iqr25	0.13	0.05	0.65	2.80	= 0.006
Zero crossings	q25	0.37	0.14	0.67	2.60	= 0.011
Zero crossings	q10	−0.26	0.10	−0.35	−2.55	= 0.012
Zero crossings	max	0.03	0.01	0.26	3.33	= 0.001
Zero crossings	var	0.01	0.00	4.12	4.41	< 0.001
Zero crossings	std	−1.33	0.34	−3.54	−3.87	< 0.001
Zero crossings	mean	−0.48	0.16	−1.23	−2.93	= 0.004
High-frequency power	median	−1.01	0.34	−1.89	−2.95	= 0.004
High-frequency power	range	0.15	0.06	0.30	2.64	= 0.010
High-frequency power	var	0.22	0.10	2.03	2.23	= 0.028
High-frequency power	std	−5.32	1.72	−2.81	−3.09	= 0.003
High-frequency power	mean	1.74	0.51	3.03	3.44	< 0.001

Note. $R^2 = 0.810, \overline{R}^2 = 0.757, F(30, 109) = 15.447, p < 0.001$.

Table 10.5 Linear regression model predicting Subjective Unit of Distress (SUD) for the Re-Living (RL) study

Feature	Parameter	B	SE (B)	β	*t*	*p*
Intercept		28.72	12.37		2.32	= 0.022
Pitch	iqr25	−0.07	0.01	−0.48	−5.22	< 0.001
Pitch	q90	0.05	0.02	0.57	2.35	= 0.020
Pitch	median	0.10	0.03	0.89	3.75	< 0.001
Pitch	var	0.00	0.00	0.39	3.90	< 0.001
Pitch	mean	−0.20	0.04	−1.64	−4.93	< 0.001
Amplitude	q75	523.83	258.34	0.57	2.03	= 0.045
Amplitude	q10	248.91	79.14	1.47	3.15	= 0.002
Amplitude	var	2304.48	964.43	1.54	2.39	= 0.018
Amplitude	std	−290.55	125.77	−2.20	−2.31	= 0.023
Power	iqr25	−1.39	0.38	−2.03	−3.62	< 0.001
Power	q25	−1.38	0.33	−2.82	−4.19	< 0.001
Power	iqr10	−0.61	0.37	−1.01	−1.67	= 0.098
Power	median	0.36	0.20	0.78	1.83	= 0.070
Power	std	5.00	1.74	3.01	2.87	= 0.005
Zero crossings	q25	0.40	0.10	0.94	3.88	< 0.001
Zero crossings	q10	−0.42	0.10	−0.74	−4.35	< 0.001
Zero crossings	median	−0.43	0.07	−1.49	−6.21	< 0.001
Zero crossings	min	0.06	0.03	0.13	2.04	= 0.043
High-frequency power	min	0.38	0.14	0.61	2.76	= 0.007
High frequency power	var	0.22	0.08	2.17	2.67	= 0.009
High-frequency power	std	−3.99	1.49	−2.27	−2.68	= 0.008
High-frequency power	mean	1.27	0.43	2.42	2.95	= 0.004

Note. $R^2 = 0.648, \overline{R}^2 = 0.584, F(22,121) = 10.118, p < 0.001$.

References

American Psychiatric Association: DSM-IVTR: Diagnostic and Statistical Manual of Mental Disorders, 4 (Text Revision) edn. American Psychiatric Publishing, Washington, DC (2000)

Banse, R., Scherer, K.R.: Acoustic profiles in vocal emotion expression. J. Personal. Soc. Psychol. **70**(3), 614–636 (1996)

Boersma, P.: Accurate short-term analysis of the fundamental frequency and the harmonics-to-noise ratio of a sampled sound. In: Proceedings of the Institute of Phonetic Sciences, vol. 17, pp. 97–110. University of Amsterdam, Amsterdam (1993)

Boersma, P.P.G., Weenink, D.J.M. Praat 4.0.4. (2006) http://www.praat.org. Last accessed on 15 Apr 2011

Cohen, A., 't Hart, J.: On the anatomy of intonation. Lingua **19**(2), 177–192 (1967)

Collier, R.: Physiological correlates of intonation patterns. J. Acoust. Soc. Am. **58**(1), 249–255 (1975)

Cousineau, D.: Confidence intervals in within-subject designs: a simpler solution to Loftus and Masson's method. Tutorials Quant. Methods Psychol. **1**(1), 42–46 (2005)

Cowie, R., Douglas-Cowie, E., Tsapatsoulis, N., Votsis, G., Kollias, S., Fellenz, W., Taylor, J.G.: Emotion recognition in human–computer interaction. IEEE Signal Process. Mag. **18**(1), 32–80 (2001)

Dash, M., Liu, H.: Feature selection for classification. Intell. Data Anal. **1**(3), 131–156 (1997)

Derksen, S., Keselman, H.J.: Backward, forward and stepwise automated subset selection algorithms: Frequency of obtaining authentic and noise variables. Br. J. Math. Stat. Psychol. **45**(2), 265–282 (1992)

Dubnowski, J., Schafer, R., Rabiner, L.: Real-time digital hardware pitch detector. IEEE Trans. Acoust. Speech Signal Process. **24**(1), 2–8 (1976)

Ehlers, A., Bisson, J., Clark, D.M., Creamer, M., Pilling, S., Richards, D., Schnurr, P.P., Turner, S., Yule, W.: Do all psychological treatments really work the same in posttraumatic stress disorder? Clin. Psychol. Rev. **30**(2), 269–276 (2010)

Geenen, R., van de Vijver, F.J.R.: A simple test of the law of initial values. Psychophysiology **30**(5), 525–530 (1993)

Greden, J.F.: The burden of recurrent depression: causes, consequences, and future prospects. J. Clin. Psychiatr. **62**(S22), 5–9 (2001)

Harrell, F.E., Jr.: Regression Modeling Strategies – with Applications to Linear Models, Logistic Regression, and Survival Analysis. Springer Series in Statistics, 1st (6th printing) edn. Springer, New York, NY (2001)

Harris, C.M., Weiss, M.R.: Pitch extraction by computer processing of high-resolution Fourier analysis data. J. Acoust. Soc. Am. **35**(3), 339–343 (1963)

Healey, E.S., Kales, A., Monroe, L.J., Bixler, E.O., Chamberlin, K., Soldatos, C.R.: Onset of insomnia: role of life-stress events. Psychosom. Med. **43**(5), 439–451 (1981)

Hilty, D.M., Marks, S.L., Urness, D., Yellowlees, P.M., Nesbitt, T.S.: Clinical and educational telepsychiatry applications: a review. Can. J. Psychiatr. **49**(1), 12–23 (2004)

Hirano, M., Ohala, J., Vennard, W.: The function of laryngeal muscles in regulating fundamental frequency and intensity of phonation. J. Speech Hear. Res. **12**(3), 616–628 (1969)

Kedem, B.: Spectral analysis and discrimination by zero-crossings. Proc. IEEE **74**(11), 1477–1493 (1986)

Kessler, R.C.: The effects of stressful life events on depression. Annu. Rev. Psychol. **48**(1), 191–214 (1997)

Khalil, E.N.: Communicating affect in news stories: the case of the lead sentence. Text Talk Interdiscip. J. Lang. Discourse Commun. Stud. **26**(3), 329–349 (2006)

Knapp, S., VandeCreek, L.: Anxiety Disorders: A Scientific Approach for Selecting the Most Effective Treatment. US: Professional Resource Press/Professional Resource Exchange, Sarasota (1994)

Kosten, T.R., Mason, J.W., Giller, E.L., Ostroff, R.B., Harkness, L.: Sustained urinary norepinephrine and epinephrine elevation in post-traumatic stress disorder. Psychoneuroendocrinology **12**(1), 13–20 (1987)

Lacey, J.I.: Somatic response patterning and stress: some revisions of activation theory, pp. 14–42. Appleton-Century-Crofts, New York, NY (1967)

Lacey, J.I., Bateman, D.E., VanLehn, R.: Autonomic response specifity: an experimental study. Psychosomat. Med. **15**(1), 8–21 TL1 (1953)

Ladd, D.R., Silverman, K.E.A., Tolkmitt, F., Bergmann, G., Scherer, K.R.: Evidence for the independent function of intonation contour type, voice quality, and *F*0 range in signalling speaker affect. J. Acoustic. Soc. Am. **78**(2), 435–444 (1985)

Lenneberg, E.H.: Biological Foundation of Language. Wiley, New York, NY (1967)

Lively, S.E., Pisoni, D.B., van Summers, W., Bernacki, R.H.: Effects of cognitive workload on speech production: acoustic analyses and perceptual consequences. J. Acoust. Soc. Am. **93**(5), 2962–2973 (1993)

Lyons, R.G.: Understanding Digital Signal Processing, 2nd edn. Prentice Hall PTR, Upper Saddle River (2004)

Marty, A.: Untersuchungen zur allgemeinen grundlegung der Grammatik und sprachphilosophie. Niemeyer, Halle/Saale (1908). Translation of title: Investigations on the general foundations of grammar and the philosophy of language

Mevissen, L., De Jongh, A.: PTSD and its treatment in people with intellectual disabilities: a review of the literature. Clin. Psychol. Rev. **30**(3), 308–316 (2010)

Morrison, D., Wang, R., De Silva, L.C.: Ensemble methods for spoken emotion recognition in call-centres. Speech Commun. **49**(2), 98–112 (2007)

Murray, I.R., Arnott, J.L.: Toward the simulation of emotion in synthetic speech: a review of the literature on human vocal emotion. J. Acoust. Soc. Am. **93**(2), 1097–1108 (1993)

Orlikoff, R.F., Baken, R.J.: Fundamental frequency modulation of the human voice by the heartbeat: preliminary results and possible mechanisms. J. Acoust. Soc. Am. **85**(2), 888–893 (1989)

Picard, R.W.: Affective Computing. MIT Press, Boston, MA (1997)

Rabiner, L., Cheng, M., Rosenberg, A., McGonegal, C.: A comparative performance study of several pitch detection algorithms. IEEE Trans. Acoust. Speech Signal Process. **24**(5), 399–418 (1976)

Rani, P., Sims, J., Brackin, R., Sarkar, N.: Online stress detection using psychophysiological signals for implicit human-robot cooperation. Robotica **20**(6), 673–685 (2002)

Ray, S.L.: Evolution of posttraumatic stress disorder and future directions. Arch. Psychiatr. Nurs. **22**(4), 217–225 (2008)

Rice, S.O.: Mathematical Analysis of Random Noise. ATI 162 535Monograph B-1589. Bell Telephone Labs, New York, NY (1952)

Rienstra, S., Hirschberg, A.: An Introduction to Acoustics, 5th edn. Extended and revised edition of IWDE 92-06 (TU/e). Eindhoven University of Technology, Eindhoven, The Netherlands (2009)

Rothkrantz, L.J.M., Wiggers, P., vanWees, J.-W.A.,, van Vark, R.J.: Voice stress analysis. Lect. Notes Comput. Sci. (Text, Speech Dialog.) **3206**, 449–456 (2004)

Rubin, D.C., Berntsen, D., Klindt Bohni, M.: A memory-based model of posttraumatic stress disorder: evaluating basic assumptions underlying the PTSD diagnosis. Psychol. Rev. **115**(4), 985–1011 (2008)

Rüscha, N., Corrigan, P.W., Powell, K., Rajah, A., Olschewski, M., Wilkniss, S., Batia, K.: A stress-coping model of mental illness stigma: II. Emotional stress responses, coping behavior and outcome. Schizophr. Res. **110**(1–3), 65–71 (2009a)

Rüscha, N., Corrigan, P.W., Wassel, A., Michaels, P., Olschewski, M., Wilkniss, S., Batia, K.: A stress-coping model of mental illness stigma: I. Predictors of cognitive stress appraisal. Schizophr. Res. **110**(1–3), 59–64 (2009b)

Scherer, K.R.: Vocal communication of emotion: a review of research paradigms. Speech Commun. **40**(1–2), 227–256 (2003)

Shimamura, T., Kobayashi, H.: Weighted autocorrelation for pitch extraction of noisy speech. IEEE Trans. Speech Audio Process. **9**(7), 727–730 (2001)

Sánchez-Meca, J., Rosa-Alcázar, A.I., Marín-Martínez, F., Gómez-Conesa, A.: Psychological treatment of panic disorder with or without agoraphobia: a meta-analysis. Clin. Psychol. Rev. **30**(1), 37–50 (2010)

Titze, I.R., Hunter, E.J.: A two-dimensional biomechanical model of vocal fold posturing. J. Acoust. Soc. Am. **121**(4), 2254–2260 (2007)

van den Broek, E.L.: Robot nannies: future or fiction? Interact. Stud. **11**(2), 274–282 (2010)

van den Broek, E.L., Janssen, J.H., Healey, J.A.,, van der Zwaag, M.D.: Prerequisites for Affective Signal Processing (ASP) – Part II. In: Fred, A., Filipe, J., Gamboa, H. (eds.) Biosignals 2010: Proceedings of the International Conference on Bio-Inspired Systems and Signal Processing, pp. 188–193. Valencia, Spain (2010a)

van den Broek, E.L., Janssen, J.H., Westerink, J.H.D.M., Healey, J.A. Prerequisites for Affective Signal Processing (ASP). In: Encarnação, P., Veloso, A. (eds.) Biosignals 2009: Proceedings of the International Conference on Bio-Inspired Systems and Signal Processing, pp. 426–433. Porto – Portugal (2009a)

van den Broek, E.L., Janssen, J.H., van der Zwaag, M.D., Healey, J.A. Prerequisits for Affective Signal Processing (ASP) – Part III. In: Fred, A., Filipe, J., Gamboa, H. (eds.) Biosignals 2010: Proceedings of the International Conference on Bio-Inspired Systems and Signal Processing, pp. 432–435. Valencia – Spain (2010b) van den Broek, E.L., Lisý, V., Janssen, J.H., Westerink, J.H.D.M., Schut, M.H., Tuinenbreijer, K.: Affective Man-Machine Interface: Unveiling Human Emotions Through Biosignals, Volume 52 of Communications in Computer and Information Science, pp. 21–47. Springer, Berlin/Heidelberg (2010c)

van den Broek, E.L., Schut, M.H., Westerink, J.H.D.M., Tuinenbreijer, K.: Unobtrusive sensing of emotions (USE). J. Ambient Intell. Smart Environ. **1**(3), 287–299 (2009b)

van den Broek, E.L., Westerink, J.H.D.M.: Considerations for emotion-aware consumer products. Appl. Ergon. **40**(6), 1055–1064 (2009)

van den Broek, E.L., van der Zwaag, M.D., Healey, J.A., Janssen, J.H., Westerink, J.H.D.M. Prerequisits for Affective Signal Processing (ASP) – Part IV. In: Kim, J., Karjalainen, P. (eds.) Proceedings of the 1st International Workshop on Bio-Inspired Human-Machine Interfaces and Healthcare Applications – B-Interface, pp. 59–66. Valencia – Spain (2010d)

Ververidis, D., Kotropoulos, C.: Emotional speech recognition: Resources, features, and methods. Speech Commun. **48**(9), 1162–1181 (2006)

Wendt, C., Petropulu, A.P. Pitch determination and speech segmentation using the discrete wavelet transform. In: IEEE International Symposium on Circuits and Systems: Connecting the World, vol. 2, pp. 45–48. IEEE Computer Society, Atlanta, GA, New York, NY (1996)

Wolpe, J.: Psychotherapy by Reciprocal Inhibition. Stanford University Press, Stanford (1958)

Zeng, Z., Pantic, M., Roisman, G.I., Huang, T.S.: A survey of affect recognition methods: audio, visual, and spontaneous expressions. IEEE Trans. Pattern Anal. Mach. Intell. **31**(1), 39–58 (2009)

Chapter 11
The Role of Design in Facilitating Multi-disciplinary Collaboration in Wearable Technology

Sharon Baurley

Abstract This chapter presents a range of methodologies that address issues around designing for the emerging area of wearable technology, based upon a 1-year research cluster, *The Emotional Wardrobe,* and a 3-year user study project, *Communication-Wear*. The process of eliciting consumer desire is very central to this in order to gain insight into the catalysts and drivers for this new genre of fashion/clothing. For this we need to elicit the dreams, aspirations and desires of people using generative techniques and prototype as *probe* methods, to provide inspiration for designers. Design for appropriation empowers people to create their own stories and meanings for an age of personalisation, enabling them to be proactive rather than reactive to technological development. This emerging design space will necessitate increased levels of collaboration between industries. But how do we work together where there isn't a history of doing so? Here we present the use of design as a way of thinking in order to manage knowledge flows, to facilitate knowledge creation, and a shared understanding.

11.1 Introduction

Fashion is a key component of consumer culture, a cultural system of making meaning, and of making meaning through what we consume; a cultural system of codes. Consumer culture is, what Raymond Williams Williams, 1981 and other writers have called, the *"bricks and mortar of everyday life"*, the music you listen to, the clothes you wear, etc. These are aspects of material culture, which we use it to map out identities for ourselves. We use fashion to define ourselves, and group ourselves into social groups and communities. What is happening now is that digital communications technologies have common attributes with fashion/clothing in terms of how they enable people to construct an identity, to be expressive, to differentiate themselves, and declare their uniqueness, which enables communication between

S. Baurley (✉)
School of Engineering and Design, Brunel University, Uxbridge, UK
e-mail: sharon.baurley@brunel.ac.uk

J. Westerink et al. (eds.), *Sensing Emotions*, Philips Research Book Series 12,
DOI 10.1007/978-90-481-3258-4_11,

people allowing them to form communities. The revolutionary growth of digital media is allowing groups and individuals to collaborate, create and share their own content and material. Sites such as YouTube allow people to express themselves, and MySpace allows people to congregate online and form communities. Teen groups experiment with language to create their own SMS codes. Mobile communications is now a part of fashion, with brands such as Prada collaborating with consumer electronics corporation LG, to develop mobile phones. The use of marketing language in Nokia's *L'Amour Collection* of phones is one of fashion, using sensory descriptors to entice consumers, enabling them to ask "which side of me shall I be"? More recently Nokia launched its Morph concept phone, in which nano materials enable it to change its shape (from a conventional handset to a bangle) and its aesthetics (downloading patterns from the phone to a handbag).

When fashion converges with ICT and materials technology, what will happen? We can't anticipate all of the end-use applications in advance because what actually happens in practice is the emergent outcome of user dynamics, e.g., texting caught service providers by surprise. If we extrapolate from what is happening in mobile and web-based communications and apply that thinking to new genres of clothing that are networked and dynamically changeable, will we see similar patterns of behaviour emerging? How can we gain prior knowledge of emergent behaviour? This type of clothing has the potential to empower people, but how can we understand what they are capable of doing? These concerns relate to new kinds of "value-added" that fulfil emotional and self-actualisation.

This chapter addresses two key issues in this space. Effective multi-disciplinary collaboration (*probing the developers*), and engaging users (*probing the users*).

This chapter describes methodologies employed in *The Emotional Wardrobe* (Baurley S., Stead L. 2007) project that attempted to use creative techniques of the designer's repertoire, as a means to facilitate collaborative working. Generative techniques were used to foster a shared understanding in brainstorm discussions, as well as to manage knowledge flows and ideas generation, again by mobilising the knowledge of individual members of the project.

The second part of this chapter describes design and design-related methodologies employed in the *Communication-Wear* project that aimed to uncover a deeper level of knowledge and understanding about user's desires, preferences,

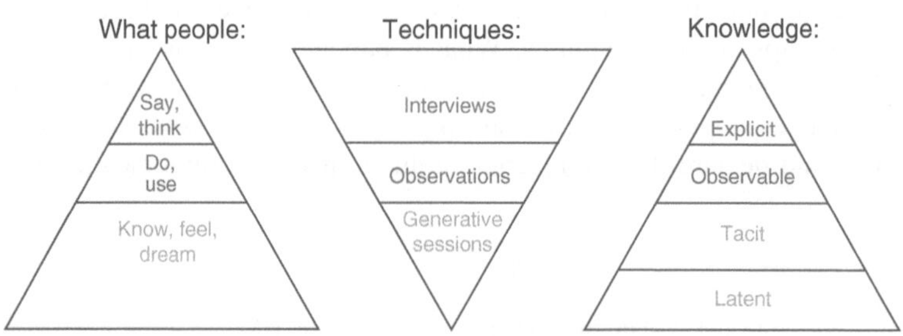

Fig. 11.1 Mobilising tacit and learned knowledge.
Source: From Sleeswick et al., 2005

beliefs, aspirations and behaviours. Generative techniques reveal a deeper level of knowledge about people's feelings, aspirations Sleeswick et al., 2005. The main focus of methodologies is the use of design and design-related methods to try to elicit consumer desire around consumer wearable technology, namely participative design, prototype as research probe and user studies. By developing prototypes using these techniques and placing them in the "in the wild", it is thought that this might help advance and inform development of products, materials and digital technologies.

11.1.1 Probing the Developers: Design as a Way of Thinking

In order to manage multi-disciplinary cooperation during workshop activity the group engaged with the concept of "thinking and knowing", i.e., knowledge and the considered application of knowledge, and experiment with creative ways of accessing their respective knowledge bases.

The aim was to use design as a way of thinking, i.e., thinking through doing, and as a means to generate knowledge and a shared understanding. Techniques included visualisation and embodiment of ideas. The success of these activities relies heavily on effective facilitation.

11.1.1.1 The Emotional Wardrobe

The central idea of the Cluster was *The Emotional Wardrobe*, in which the conventions and cultures of fashion, as an expressive, emotional and communicative medium, are extended by integrating computer intelligence and digital communications. The main themes of the *The Emotional Wardrobe* were:

Emotional Connection. This is about gaining an understanding of how we create meaning through what we consume. By enabling individuals to build their

Fig. 11.2 The group re-modelling 2nd hand garments

own stories using personally relevant information including moods, interests, history, geography and ethical concerns, we can start to gain insight into how people form emotional bonds with objects.

Human Connectedness. This is about broadening the range of expressive communication channels of clothing. By bringing sensor and network technology to the body, new forms of communication and interaction could be enabled.

Customisation and Creativity. This is about enhancing people's expressive and creative possibilities. It has been mooted that in the future creativity will no longer be the preserve of the "creative class". It will be imperative to educate and develop people who think in lateral and creative ways. Clothing is an expressive medium; it facilitates individualistic expression, allowing individuals to differentiate themselves and to declare their uniqueness. Clothing aesthetics that can be dynamically personalised could encourage new ways of creative thinking through aesthetic, informative, cultural and gaming explorations.

Open Forum Workshop: What and How to Explore

The *Open Forum* workshop was a facilitated workshop, used to formulate what and how we would explore in the forthcoming workshops during the 12 months duration of the project. The first two workshops aimed to get everyone thinking about fashion and their own clothing.

"Bring and Tell"

Project members brought in a favourite garment to discuss issues around: Whether it evokes any particular emotions or memories; whether it has a story; whether it communicates their identity; as well as issues around when they wear it, and how it makes them feel.

"Scrapbox Challenge"

To continue the exploration of personal choice during the "*Scrapbox Challenge*" we each selected a garment that we disliked or had a negative emotional reaction to, from a selection of second-hand clothes. We evaluated its negative features in terms of its mis-match to our identity or interests. We explored ways in which we could reconstruct the garment to make it acceptable or to make it "fit" our identity, lifestyle, and preferences.

Using other garments, fabrics and haberdashery we recreated the garment: We transplanted parts of other garments, added surface details or reshaped the garment cutting and added pieces of fabric.

In small groups we discussed the changes made and why they were made; how participants made their decisions; how the meaning created by the maker was perceived by others in the group. We also discussed the experience of deconstructing and reconstructing, and people from non-design backgrounds found the task.

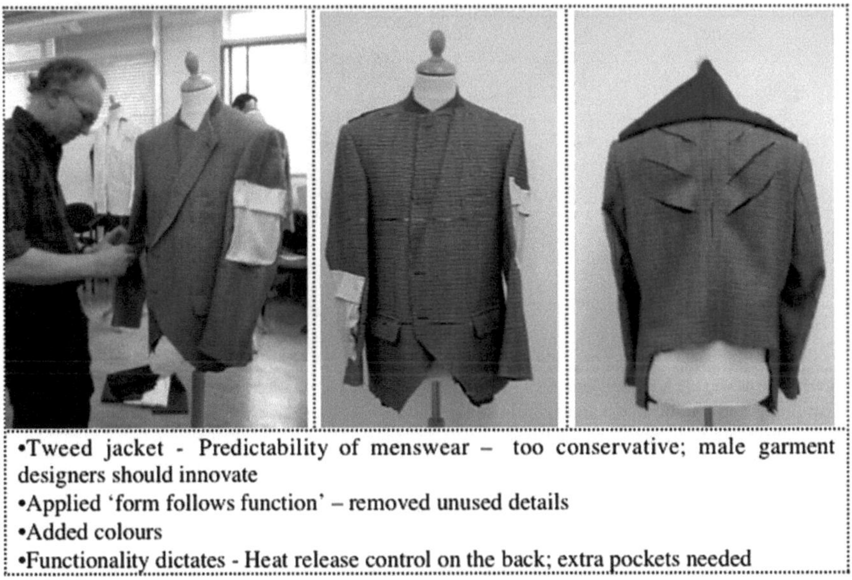

Fig. 11.3 A garment re-construction by a member of the project team

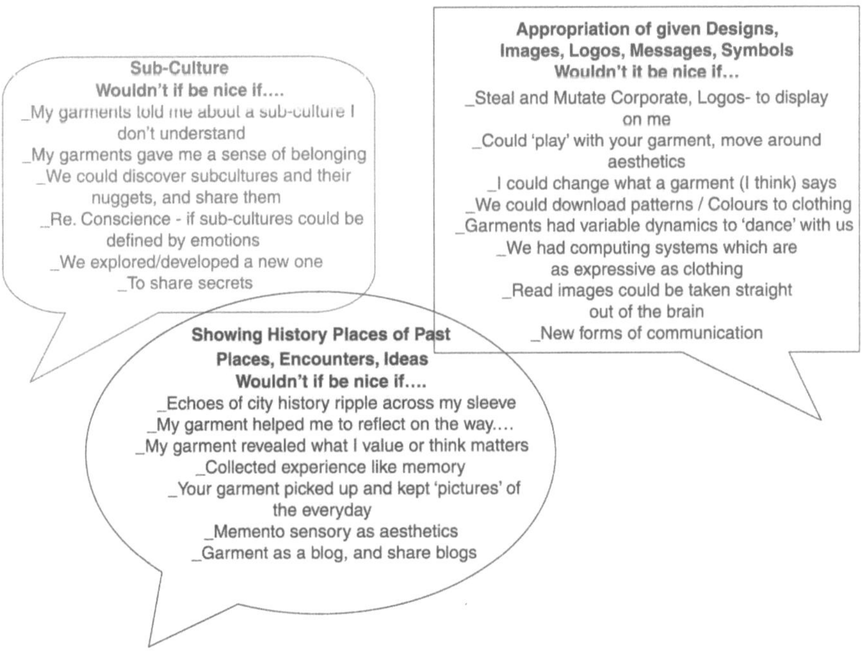

Fig. 11.4 "Brain-drawing" bubbles

Scene 1 Scene 2 Scene 3

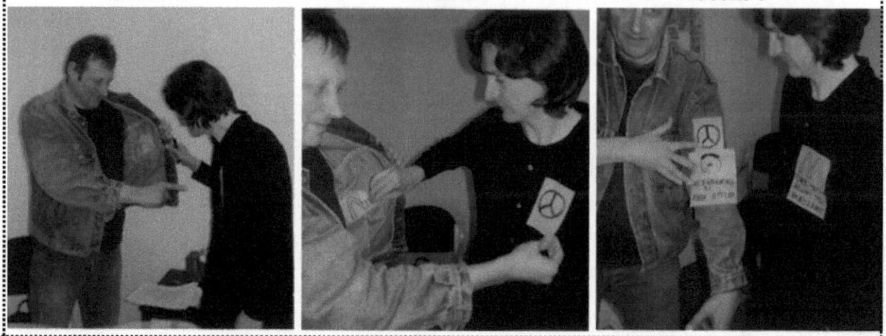

- Two groups see each others' images - swapping, adjust by tactile interaction on jacket.
- Swapping and trading like badges – gifts in a culture.
- Copying, grabbing, transferring images.
- Original images grabbed from camera phone, web.
- Ability to hack or extend markings and symbols.
- Textile as drawing surface, and drag image to the best place to show it.

Fig. 11.5 Trading and "warping" logos scenario

"Daydreaming" and "Brain Drawing"

We engaged in some facilitated "daydreaming", thinking up "wouldn't it be nice if..." scenarios/ideas, based around the three themes. Using "*Brain Drawing*" techniques we condensed the "wouldn't it be nice if" scenarios into problem statements to be explored in the *Explore* workshop.

Explore Workshop: Scenario-Building

"Role Play" or Bodystorming

We explored the problem statements through scenario-building, for which we took a divergent approach. As it is a given that garments are worn most of time, the aim was to concentrate on the creative exploration of the interactions, environments, situations, problems and limitations encountered in everyday life, which were mapped out in scenarios. The purpose of these scenarios was to locate a meaningful time and place for a technological intervention and the interesting questions and issues that are posed. We wanted to identify the research issues and find out how we could conceptualise and explore them.

The "Customisation and Creativity" problem statement was: How to collect and share images, logos and graphics? What they are, when we would want to collect/share and exchange them, and how the wearer could adapt and control the activity?

We used a range of generative techniques to conceptualise and explore the scenarios. We explored "a day in the life of" using role-play and photographed them. Ideas can be generated and tested quickly using this method, as it mobilizes intuitive

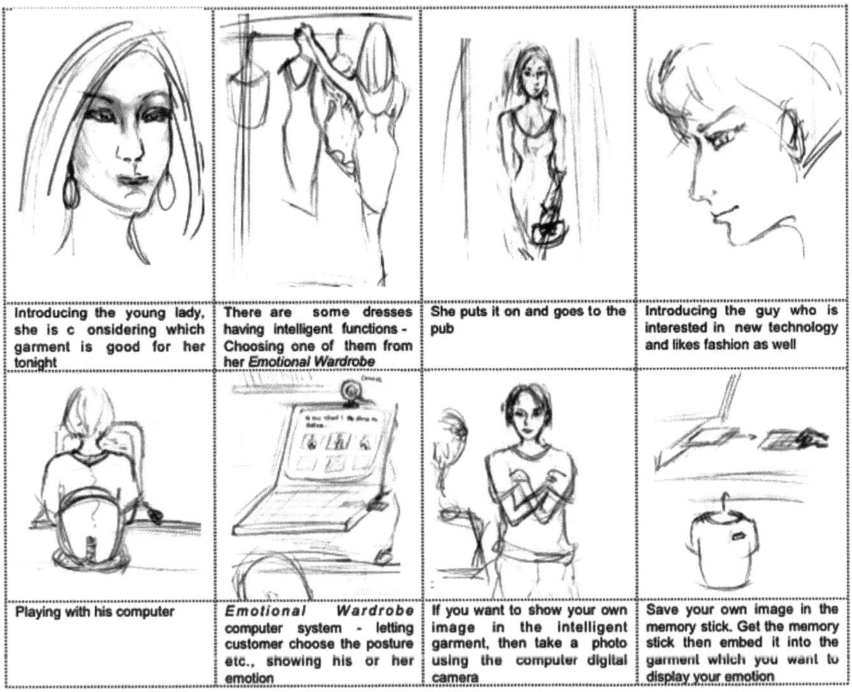

Fig. 11.6 Smart garment storyboard sketch (illustrations by K.F. You)

actions and reactions. We arranged the photos into a time line. We annotated the photos with key moments of interaction to elaborate the scenario into a storyboard. We looked at the scenario from a close-up (personal, emotional) level, medium range shot (supporting people and actions), outside the image (context). We looked for moments where "customization and creativity" could play a part.

We then went onto consider the role of technology. We looked at whether the garments could be enhanced to change, extend, and facilitate the interactions; and whether they would "behave" on command instinctively, or be triggered to react, and what would be the trigger.

Create Workshop: "Visualisation" and Embodiment of Ideas

In the *Create* workshop we elaborated the scenarios through making. The "Customisation and Creativity" group worked with a disco/dance club scenario. The idea is that body movements used by the dancers are specific to a sub-culture, and are used to control the exchange of images between people to display on clothing.

We used visualisation techniques to tell a story, which we would present to a user group in the final workshop, *Observe*. Initial ideas were sketched by animation students as they were discussed, in order to promote a shared understanding amongst

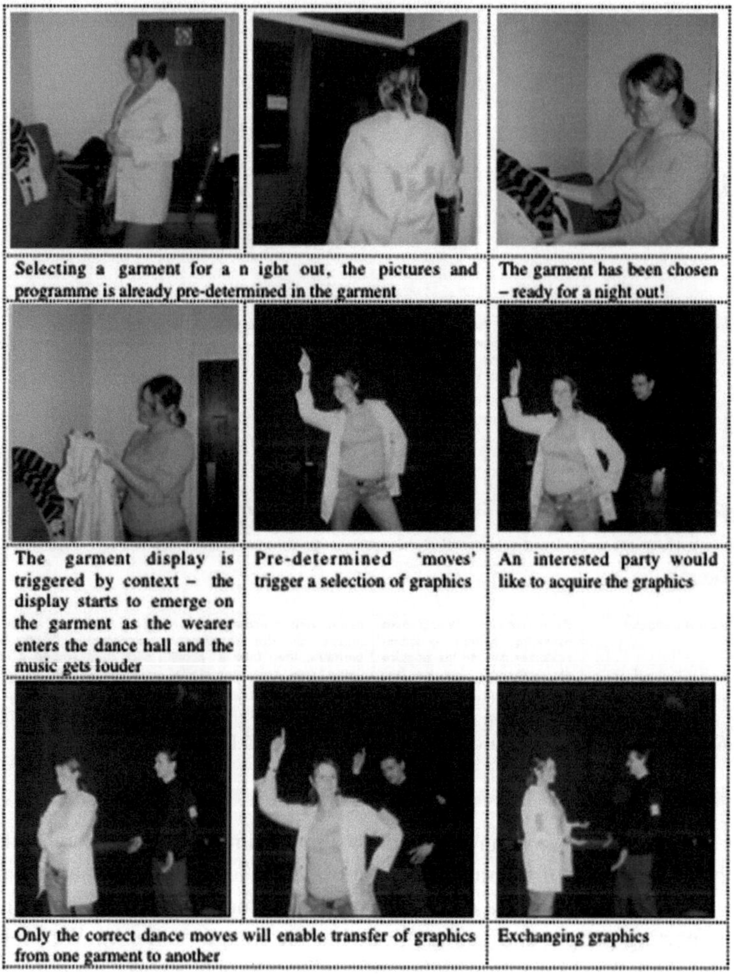

Fig. 11.7 Storyboard of actors acting-out the dancehall scenario

the project group. These sketches were used as a basis for mocking-up garment prototypes.

The scenarios were then "role-played" by actors as a way of concretising the stories, which were filmed and photographed. Working with the actors was especially helpful as they were able to input their experience, natural reactions and gestures.

Observe Workshop: Eliciting Feedback

The Observe workshop was about eliciting feedback from a teen group on the scenarios and stories embodied in garment prototypes and visualisations.

Fig. 11.8 Favourite outfit by a user study participant

We got participants to generate stories by expressing the relationships they have with their clothing, e.g., favourite outfit, the kind of outfit they would wear for a particular occasion, and then to give that outfit a persona, and how that persona would behave in a given situation. Collages were created from images, text and colour. The storyboards comprised tear sheets from magazines, images and notation that are figurative and abstract representations of their personal experiences, developed into stories. Such techniques may help to identify beliefs and personal experiences among age groups that may not respond well to the more conservative questionnaires and interviews often used in research. These stories can provide inspiration for products.

Taking the idea of trading images and logos we developed prototype mock-ups as "probes" to demonstrate input and output mechanisms of the technology to the user study participants, to give them an idea of what these products might look and feel like. We also presented the storyboards to participants. Again by embodying ideas we can inspire participants to reveal a deeper level of knowledge in a facilitated brainstorm, where subjects were asked to provide responses to questions on the "Customisation and Creativity" thematic area: What if you could adapt your clothing, and use others' clothing to do that? If you saw a pattern or logo on another's shirt, would you like to take it and put it onto yours? A selection of the responses can be seen in Fig. 11.9.

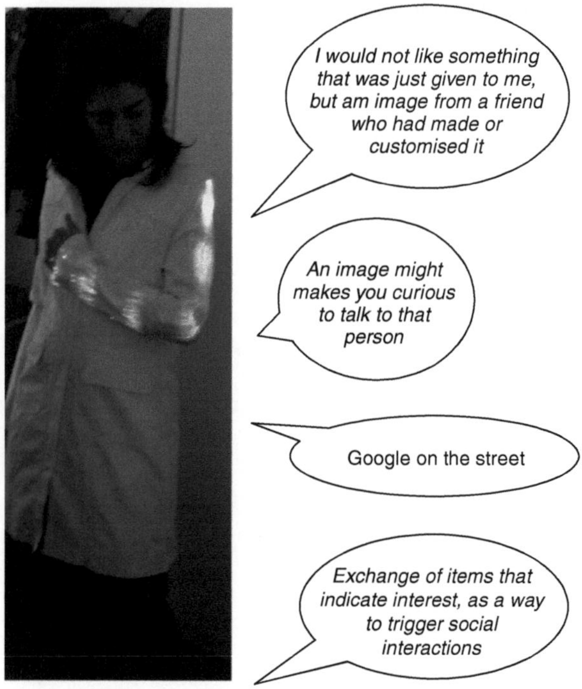

Fig. 11.9 Prototype as design probe with responses from users

11.2 Probing the Users: Communication-Wear

The notion of using prototypes as probes to enable users to provide inspiration for product development was taken further in *Communication-Wear.*

We communicate and express through the well-established "media" of clothing, spoken word, bodily and facial gestures, written and electronic words and codes. What if people could compose their own languages, codes or moods to communicate and express, using smart materials? The number of possibilities and combinations of language that smart materials might yield could be vast, i.e., colour, pattern, tactile quality, shape. So how can developers design for, and facilitate, this level of appropriation?

Communication-Wear Baurley et al., 2007, 2009 is a wearable tech(nology) clothing concept that augments the mobile phone by enabling expressive messages to be exchanged remotely, through conveying a sense of touch, and presence. Using smart textiles in garment prototypes as part of an on-going iterative co-design process, we endeavoured to mobilise participants' tacit knowledge in order to gauge user perceptions on touch communication in user trials. The aim of this study was to determine whether established sensory associations people have with the tactile qualities of textiles could be used as signs and metaphors for experiences, moods, social interactions and gestures, related to interpersonal touch.

Fig. 11.10 Probe experiments in the watershed media centre, Bristol

A series of design experiments were conducted during the course of the project that explored how users might appropriate the sensory properties of a future world of smart materials, as signs and metaphors for experiences, moods, social interactions, identity. The process of designing touch actuators involved participative design, which mobilises tacit knowledge from people about their experiences, preferences, and associations, combined with the designers' knowledge and experience of designing for the human condition. These experiments comprised iterative generative techniques, namely mock-ups and prototypes of garments as design probes, deployed during user studies, which were used to explore associations participants have between personal expression and communication, and colours, shapes, patterns, and tactile qualities.

During the user studies pairs of participants spent time exchanging touch messages, using SMS or gesture or physiological response. The prototypes were used as probes during the interviews that followed. The SMS platform was developed by Vodafone. The user studies were conducted in collaboration with HP Labs.

Textiles have a range of tactile qualities, which textile and fashion designers have always exploited as part of their design method to engineer a look, concept, mood. There are well-established descriptors for the sensory and *hand* qualities of textiles used in the fashion and textile industry Kawabata, 1980 as part of the process to select a textile for a particular clothing application. There is an industry-standard set of bi-polar attributes for fabric hand, e.g., smooth-rough, soft-crisp, cool-warm, delicate-coarse, hard-soft. These descriptors along with other references, such as colour, shape, pattern, are used by fashion and textile designers as a legitimate design method. These collections can become trends or genres (depending on consumer take-up), and become a part of consumer culture, much like languages are doing in social networking. For example, a designer would start devising a design collection with a storyboard that communicated its *mood* on a visual and tactile level. If a key component of the collection was a *warm* mood, the designer would

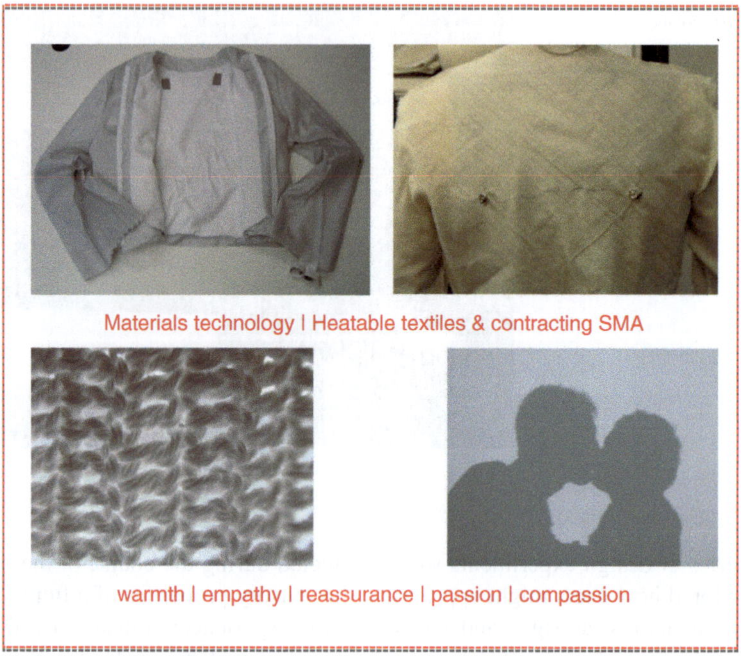

Fig. 11.11 Results storyboard

compose swatches of fabric that were warm to the touch and had a warm aesthetic, together with a warm colour palette, as well as contextual images that communicated a sense of *warm*. The selection of these swatches and images would be informed by established cultural understandings of them, which the designer understands. This was the design process employed to design the touch actuators. The designs were refined with each user study, and the final iterations are presented here.

The range of textile actuators included shape-changing, and heat and light-emitting textiles, as well as electronic textile sensors, namely touch, mechanical stretch (gesture), and GSR. All electronic textiles used in this prototype were produced using a weaving process and silver-coated nylon yarn. The placement of these actuators is informed by Argyle's "vocabulary of touch" Argyle, 2001, which is based on research into interpersonal touch points on the body.

Touch actuators included heatable textiles, textiles that change from being cool to warm upon receipt of a touch communication. A fabric that has a warm handle is generally understood to have *comforting* associations; synonyms for *warm* include having or displaying warmth or affection, passionate, friendly and responsive, caring, fond, tender. The heat pads were located on the upper back of the jacket. The results suggested that a warm tactile sensation delivered through heatable textiles evoked a sense of reassurance and empathy.

A tactile actuator that attempted to simulate a *stroking* sensation on the arms was engineered using shape memory alloy wire together with a pleated fabric insert.

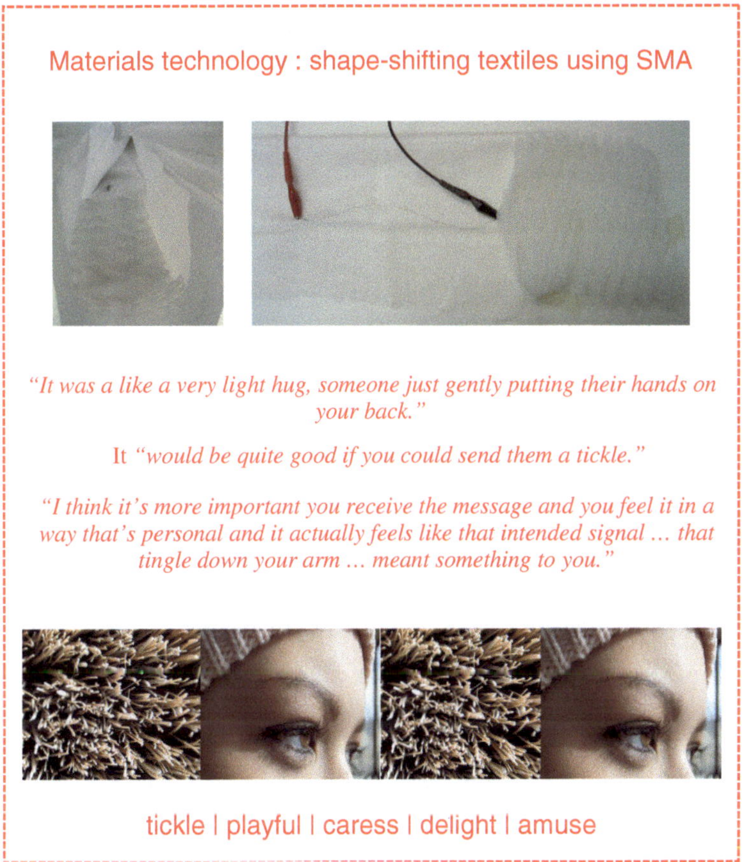

Fig. 11.12 Results storyboard

This pleated insert was located on the inside of the lower part of the sleeve, so that it would slide against the topside of the lower part of the arm. A silk-like fabric was chosen that would deliver a smooth, light tickling sensation. The results suggested that a fabric that moved against the skin using shape-shifting textiles generated a tickling sensation, and evoked thoughts of fun and playfulness.

Woven fibre-optic fibres were engineered into the garment on the underside of the sleeves. Physiological arousal, as detected by the GSR sensors, was relayed to the recipient by light being emitted from the fibre-optic section. The purpose of the GSR was to see whether receiving touch messages aroused the participants. The results suggested that lustrous or light-emitting fabrics evoked feelings of radiance and happiness, and having a glow. There was also great excitement at seeing the emotional response of the study partner visualized by the light-emitting material.

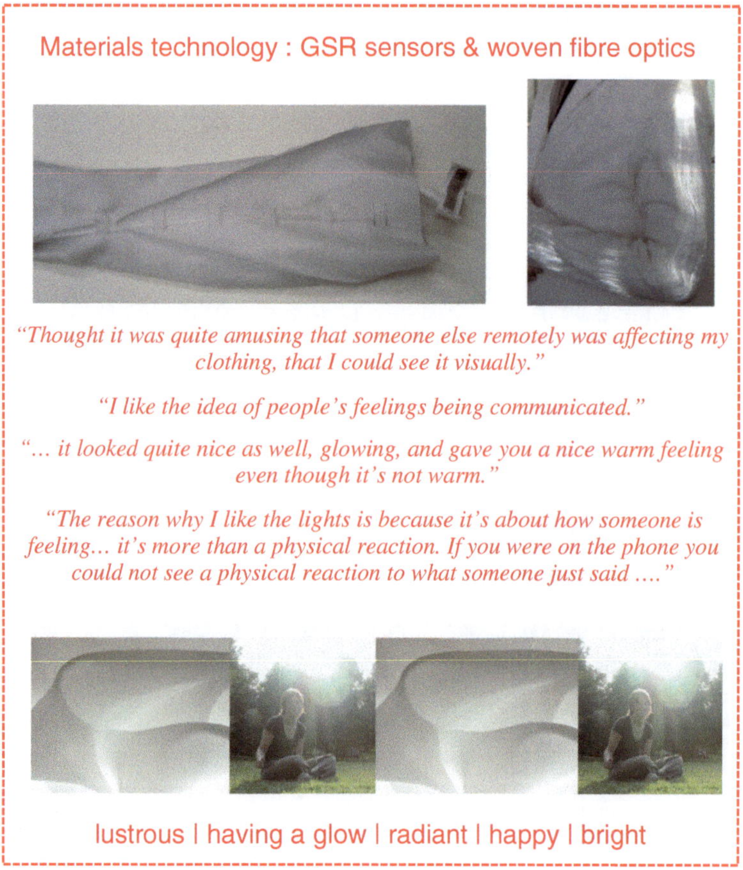

Fig. 11.13 Results storyboard

11.3 Conclusion

The Emotional Wardrobe: Multi-disciplinary working requires an effective means with which to communicate with each other. Using design to embody ideas and bring them to life promotes a shared understanding, and in so doing mobilizes the appropriate pieces of knowledge from each member of the team needed to advance an idea. These methods enable new product areas to be scoped quickly, without the risks associated with commercial product development.

Communication-Wear: As communication is personal, and just like writing, there is a need for a universal language of sensations that people can configure to make multiple meanings with. We discerned how the design process could facilitate working collaboratively with users in the development process. By using design to embody ideas, designers and developers can inspire users, so that users can inspire designers.

Acknowledgements Acknowledgements are due to the participants in the *Communication-Wear* project, funded by The Arts and Humanities Research Council's Fellowships in the Creative and Performing Arts scheme, UK: Philippa Brock of Central Saint Martins College of Art & Design, Erik Geelhoed of Hewlett-Packard Labs, Andrew Moore, a consultant, Phil Gosset of Vodafone Future Studies. Acknowledgements are due to the participants in *The Emotional Wardrobe* project, funded by The Engineering and Physical Sciences Research Council/Arts and Humanities Research Council *Designing for the twenty-first Century* scheme, UK: Martin Woolley and Stephen Scrivener of Central Saint Martins College of Art and Design, Erik Geelhoed of Hewlett-Packard Labs, Phil Gosset of Vodafone Future Studies, Matthew Chalmers of the Department of Computing Science at University of Glasgow, Peter Excell of the School of Informatics at University of Bradford, Joan Farrer of the School of Fashion and Textile Design at Royal College of Art, Jeremy Pitt and Petar Goulev of the Department of Electronic and Electrical Engineering at Imperial College London, and Christian Heath of the Management Centre at King's College London. Thanks are also due to Remko van der Lugt of the Technical University of Delft, Netherlands, and Stijn Ossevort.

References

Argyle, M.: Bodily Communication. Routledge, London (2001)

Baurley, S., Brock, P., Geelhoed, E., Moore, A.: Communication-Wear: User Feedback as Part of a Co-design Process (2). In Oakley, Ian and Brewster, Stephen A. (Eds.): Proceedings of 2nd International Workshop on Haptic and Audio Interaction Design. Lecture Notes in Computer Science (LNCS), 56–68. Springer, Heidelberg (2007)

Baurley, S., Brock, P., Geelhoed, E., Moore, A.: Communication-wear: user feedback as part of a co-design process. In: Gebhardt, J., Greif, H., Raycheva, L., Lobet-Maris, C., Lasen, A. (eds.) Experiencing Broadband Society, 169–185. Peter Lang, Berlin (2009)

Baurley, S., Stead, L.: The emotional wardrobe. In: Inns, T. (ed.) Designing for the 21st Century: Interdisciplinary Questions and Insights, 273–298. Gower Ashgate, London (2007)

Kawabata, S.: The Standardization and Analysis of Hand Evaluation. Textile Machinery Society of Japan, Osaka (1980)

Sleeswick Visser, F., Stappers, P.J., van der Lugt, R., Sanders, E.B.N.: Contextmapping: Experiences from Practice. CoDesign, Taylor Francis, Abingdon (2005)

Williams, R.: Culture. Fontana New Sociology Series, Glasgow (1981)

MIX
Papier aus verantwortungsvollen Quellen
Paper from responsible sources
FSC® C105338

If you have any concerns about our products,
you can contact us on
ProductSafety@springernature.com

In case Publisher is established outside the EU,
the EU authorized representative is:
Springer Nature Customer Service Center GmbH
Europaplatz 3, 69115 Heidelberg, Germany

Printed by Libri Plureos GmbH
in Hamburg, Germany